Springer Tracts in Natural Philosophy

Volume 30

Edited by C. Truesdell

Springer Tracts in Natural Philosophy

Vol. 1 Gundersen: Linearized Analysis of One-Dimensional Magnetohydrodynamic Flows.
With 10 figures. X, 119 pages. 1964.

Vol. 2 Walter: Differential- und Integral-Ungleichungen und ihre Anwendung bei Abschätzungs- und Eindeutigkeitsproblemen
Mit 18 Abbildungen. XIV, 269 Seiten, 1964.

Vol. 3 Gaier: Konstruktive Methoden der konformen Abbildung
Mit 20 Abbildungen und 28 Tabellen. XIV, 294 Seiten. 1964.

Vol. 4 Meinardus; Approximation von Funktionen und ihre numerische Behandlung
Mit 21 Abbildungen. VIII, 180 Seiten, 1964.

Vol. 5 Coleman, Markovitz, Noll: Viscometric Flows of Non-Newtonian Fluids. Theory and Experiment
With 37 figures. XII, 130 pages. 1966.

Vol. 6 Eckhaus: Studies in Non-Linear Stability Theory
With 12 figures. VIII, 117 pages. 1965.

Vol. 7 Leimanis: The General Problem of the Motion of Coupled Rigid Bodies About a Fixed Point
With 66 figures, XVI, 337 pages. 1965.

Vol. 8 Roseau: Vibrations non linéaires et théorie de la stabilité
Avec 7 figures. XII, 254 pages. 1966.

Vol. 9 Brown: Magnetoelastic Interactions
With 13 figures. VIII, 155 pages. 1966.

Vol. 10 Bunge: Foundations of Physics
With 5 figures. XII, 312 pages. 1967.

Vol. 11 Lavrentiev: Some Improperly Posed Problems of Mathematical Physics
With 1 figure. VIII, 72 pages. 1967.

Vol. 12 Kronmüller: Nachwirkung in Ferromagnetika
Mit 92 Abbildungen. XIV, 329 Seiten. 1968.

Vol. 13 Meinardus: Approximation of Functions: Theory and Numerical Methods
With 21 figures. VIII, 198 pages. 1967.

Vol. 14 Bell: The Physics of Large Deformation of Crystalline Solids
With 166 figures. X, 253 pages. 1968.

Vol. 15 Buchholz: The Confluent Hypergeometric Function with Special Emphasis on its Applications
XVIII, 238 pages. 1969.

Vol. 16 Slepian: Mathematical Foundations of Network Analysis
XI, 195 pages. 1968.

Vol. 17 Gavalas: Nonlinear Differential Equations of Chemically Reacting Systems
With 10 figures. IX, 107 pages. 1968.

Vol. 18 Marti: Introduction to the Theory of Bases
XII, 149 pages. 1969.

William Alan Day

Heat Conduction Within Linear Thermoelasticity

Springer-Verlag
New York Berlin Heidelberg Tokyo

William Alan Day
Mathematical Institute
University of Oxford
24–29 St. Giles
Oxford OX1 3LB
England

AMS Classifications: 73B30, 73C25, 80A20

Library of Congress Cataloging in Publication Data
Day, William Alan
Heat conduction within linear thermoelasticity.
(Springer tracts in natural philosophy; v. 30)
Bibliography: p.
Includes index.
1. Thermoelasticity. 2. Heat equation—Numerical solutions. I. Title. II. Series.
QA933.D38 1985 536'.41'015153 85-11455

Typeset by Composition House Ltd., Salisbury, England.
Printed and bound by Graphischer Betrieb Konrad Triltsch, Würzburg, West Germany.
Printed in West Germany.

9 8 7 6 5 4 3 2 1

ISBN 0-387-96156-9 Springer-Verlag New York Berlin Heidelberg Tokyo
ISBN 3-540-96156-9 Springer-Verlag Berlin Heidelberg New York Tokyo

Contents

Introduction

J-B. J. FOURIER's immensely influential treatise *Théorie Analytique de la Chaleur* [21], and the subsequent developments and refinements of FOURIER's ideas and methods at the hands of many authors, provide a highly successful theory of heat conduction.

According to that theory, the growth or decay of the temperature θ in a conducting body is governed by the *heat equation*, that is, by the parabolic partial differential equation

$$\frac{\partial^2\theta}{\partial x^2}+\frac{\partial^2\theta}{\partial y^2}+\frac{\partial^2\theta}{\partial z^2}=\frac{\partial\theta}{\partial t}.$$

Such has been the influence of FOURIER's theory, which must forever remain the *classical* theory in that it sets the standard against which all other theories are to be measured, that the mathematical investigation of heat conduction has come to be regarded as being almost identical† with the study of the heat equation, and the reader will not need to be reminded that intensive analytical study has

† But not entirely; witness, for example, those theories which would replace the heat equation by an equation which implies a finite speed of propagation for the temperature. The reader is referred to the article [9] of COLEMAN, FABRIZIO, and OWEN for the derivation of such an equation from modern Continuum Thermodynamics and for references to earlier work in this direction.

amply demonstrated that the heat equation enjoys many properties of great interest and elegance.†

The arguments upon which the derivation of the heat equation is based presume the conducting body to be *rigid*, and, thus, they ignore any possible interaction between thermal effects and mechanical effects. It is the purpose of this tract to suggest that insight into the nature of thermomechanical interaction can be obtained by studying what is a very restricted subject indeed, namely heat conduction according to the one-dimensional version of the equations of linear thermoelasticity for a homogeneous and isotropic body. These equations constitute the simplest generalization of the heat equation which incorporates the effect of thermomechanical coupling and the effect of inertia. At all points we shall attempt to point out both the contrasts and the similarities between the heat equation and the thermoelastic equations.

The tract does not pretend to be a systematic or complete account of linear thermoelasticity, and, indeed, it is difficult to see how such an account could be written at the present time, for the investigation of thermomechanical interaction is not sufficiently far advanced. In saying this I intend no disparagement of such standard works as those of BOLEY and WEINER [4], CARLSON [6], CHADWICK [8], or NOWACKI [29], but I would point out that what is known of the implications of the coupled dynamic theory is slight by comparison with what is known of the implications of the uncoupled or equilibrium theories.

† Detailed accounts of the heat equation are to be found in the treatises of CANNON [5], CARSLAW and JAEGER [7], and WIDDER [32].

CHAPTER 1

Preliminaries

§1.1. One-dimensional Linear Thermoelasticity

We confine our attention to a homogeneous and isotropic body which is bounded by a pair of parallel plane faces $x = 0$ and $x = l$, where x is a Cartesian coordinate. We suppose the body to have a uniform stress-free reference state in which the absolute temperature is a positive constant θ_0 (independent of x); we call θ_0 the *reference temperature*. We suppose too that there is no external supply of heat to the body, nor any body force, and that the displacement vector, measured from the configuration of the reference state, remains parallel to the x-axis throughout the body and at every time t.

In these circumstances, the equations of linear thermoelasticity, which can be derived as is done in CARLSON's *Handbuch* article [6], reduce to an *equation of energy*

$$k \frac{\partial^2 \theta}{\partial x^2} = c \frac{\partial \theta}{\partial t} + \theta_0 \alpha(3\lambda + 2\mu) \frac{\partial^2 u}{\partial x \, \partial t},$$

and an *equation of motion*

$$(\lambda + 2\mu) \frac{\partial^2 u}{\partial x^2} = \alpha(3\lambda + 2\mu) \frac{\partial \theta}{\partial x} + \rho \frac{\partial^2 u}{\partial t^2},$$

which connect the partial derivatives of the *absolute temperature* θ and of the *displacement* u.

Apart from the reference temperature θ_0, the constants which appear in the equations are: the *thermal conductivity* k, the *specific heat* c, the *coefficient of thermal expansion* α, the *elastic moduli* λ and μ, and the *mass density* ρ.

If we introduce, in addition, the *heat flux*

$$q = -k\frac{\partial\theta}{\partial x},$$

the *entropy*

$$\eta = \frac{c}{\theta_0}(\theta - \theta_0) + \alpha(3\lambda + 2\mu)\frac{\partial u}{\partial x},$$

and the *stress*

$$\sigma = (\lambda + 2\mu)\frac{\partial u}{\partial x} - \alpha(3\lambda + 2\mu)(\theta - \theta_0),$$

we can rewrite the equation of energy as

$$-\frac{\partial q}{\partial x} = \theta_0\frac{\partial\eta}{\partial t},$$

and we can rewrite the equation of motion as

$$\frac{\partial\sigma}{\partial x} = \rho\frac{\partial^2 u}{\partial t^2}.$$

The partial derivatives

$$\frac{\partial u}{\partial x},\quad \frac{\partial^2 u}{\partial x\,\partial t},\quad \frac{\partial u}{\partial t},\quad \frac{\partial^2 u}{\partial t^2},$$

of the displacement are just the *strain*, the *strain rate*, the *velocity*, and the *acceleration*, respectively.

It is convenient to reduce the excessive number of constants by making the change of variables

$$x \to \frac{x}{l}, \qquad t \to \frac{kt}{cl^2},$$

$$\theta \to \frac{\theta - \theta_0}{\theta_0}, \qquad u \to \frac{u}{l}\sqrt{\frac{\lambda + 2\mu}{\theta_0 c}}.$$

The result of this change is to replace the interval $[0, l]$ by the unit interval $[0, 1]$, to replace the reference temperature by $\theta = 0$, and to transform the equation of energy and the equation of motion into the *thermoelastic equations*

$$\frac{\partial^2 \theta}{\partial x^2} = \frac{\partial \theta}{\partial t} + \sqrt{a} \cdot \frac{\partial^2 u}{\partial x\, \partial t}, \tag{1.1.1}$$

$$\frac{\partial^2 u}{\partial x^2} = \sqrt{a} \cdot \frac{\partial \theta}{\partial x} + b\frac{\partial^2 u}{\partial t^2}, \tag{1.1.2}$$

where a and b are the positive and dimensionless constants

$$a = \frac{\theta_0 \alpha^2 (3\lambda + 2\mu)^2}{c(\lambda + 2\mu)},$$

$$b = \frac{k^2 \rho}{c^2(\lambda + 2\mu) l^2}.$$

Of the two constants, the *coupling constant* a, which is usually small by comparison with unity, is independent of the thickness l and is a measure of the coupling between thermal and mechanical effects, while b, the *inertial constant*, decreases to zero as the thickness increases to infinity and is a measure of the effect of inertia.

In the light of these changes of variables we also modify the definitions of the heat flux, the entropy, and the stress, to read as

$$q = -\frac{\partial \theta}{\partial x},$$

$$\eta = \theta + \sqrt{a} \cdot \frac{\partial u}{\partial x},$$

$$\sigma = \frac{\partial u}{\partial x} - \sqrt{a} \cdot \theta,$$

respectively.

Since it is our intention to draw comparisons between the thermoelastic theory and the classical theory for a rigid conductor, *we shall suppose throughout that the faces of the body are clamped in an attempt to hold the body as rigid as can be, that is to say*

$$u|_{x=0} = u|_{x=1} = 0. \tag{1.1.3}$$

If we set $a = b = 0$ in the thermoelastic equations, and appeal to (1.1.3) as well, we recover the equations of the one-dimensional version of FOURIER's theory, according to which the temperature is a solution of the *heat equation*

$$\frac{\partial^2 \theta}{\partial x^2} = \frac{\partial \theta}{\partial t}$$

and the displacement vanishes identically.

On the other hand, if we suppose a and b to be positive and we eliminate the displacement between the thermoelastic equations, we find that the temperature is a solution not of the heat equation but of the fourth-order equation†

$$\frac{\partial^4 \theta}{\partial x^4} + b\frac{\partial^3 \theta}{\partial t^3} = (1 + a)\frac{\partial^3 \theta}{\partial x^2\, \partial t} + b\frac{\partial^4 \theta}{\partial x^2\, \partial t^2}. \tag{1.1.4}$$

The comparative complexity of this equation, as against that of the heat equation, warns of the difficulties that confront us when we attempt to discuss heat conduction within even a linear theory of thermoelasticity.

We pause to note that, except in Chapter 3, we shall be concerned exclusively with solutions θ, u of the thermoelastic equations which are defined and continuous in the half-strip $[0, 1] \times [0, \infty)$ and possess an appropriate number of continuous derivatives in $[0, 1] \times (0, \infty)$. In Chapter 3, where we consider trigonometric solutions, θ and u are defined on the entire strip $[0, 1] \times (-\infty, \infty)$ and have continuous derivatives of all orders.

§1.2. An Energy Integral

We conclude this brief account of preliminary matters by recording a formula which will play an important role in some of our arguments, notably those of Chapters 4 and 5. In order to state it we introduce the *total energy density*

$$e = \frac{1}{2}\left(\theta^2 + \left(\frac{\partial u}{\partial x}\right)^2 + b\left(\frac{\partial u}{\partial t}\right)^2\right)$$

and the *total energy*

$$E = \int_0^1 e\, dx,$$

† Equation (1.1.4) is the one-dimensional, homogeneous, and isotropic version of the equation that CARLSON [6] calls the *temperature equation*.

each being nonnegative. As we can readily verify, the thermoelastic equations and the definitions of the heat flux q and the stress σ imply the identity

$$\frac{\partial e}{\partial t} + q^2 = \frac{\partial}{\partial x}\left(-\theta q + \sigma \frac{\partial u}{\partial t}\right),$$

and when we integrate with respect to x and appeal to the boundary conditions (1.1.3) we arrive at

Theorem 1. *If θ and u are C^2 in $[0, 1] \times (0, \infty)$, the derivative E' of the total energy exists on $(0, \infty)$ and is given by*

$$E' + \int_0^1 q^2\,dx = -(\theta q)|_{x=1} + (\theta q)|_{x=0}. \tag{1.2.1}$$

Equation (1.2.1) is a very special case of what CARLSON [6] calls the Theorem of Power and Energy. That theorem can be used, in a standard way, to establish uniqueness for the boundary and initial value problem of dynamic linear thermoelasticity.

CHAPTER 2

The Coupled and Quasi-static Approximation

§2.1. An Integro-differential Equation

This chapter is given over to an investigation of the effect of thermomechanical coupling in isolation from the effect of inertia. To this end we make the *coupled and quasi-static approximation* which entails that the coupling constant a be positive while the inertial constant b is set equal to zero. Thus, we retain the first of the thermoelastic equations (1.1.1) but replace (1.1.2) by the approximate equation

$$\frac{\partial^2 u}{\partial x^2} = \sqrt{a} \cdot \frac{\partial \theta}{\partial x}. \tag{2.1.1}$$

In the light of the boundary conditions (1.1.3) the strain $\partial u/\partial x$ must satisfy the condition

$$\int_0^1 \frac{\partial u}{\partial x}\, dx = 0,$$

and if we integrate (2.1.1) with respect to x and appeal to this condition we can express the strain in terms of the temperature by way of the equation

$$\frac{\partial u}{\partial x}(x, t) = \sqrt{a} \cdot \theta(x, t) - \sqrt{a} \cdot \int_0^1 \theta(y, t)\, dy. \tag{2.1.2}$$

When we differentiate with respect to t and substitute for the strain rate in (1.1.1) we conclude that the temperature is a solution of the equation

$$\frac{\partial^2\theta}{\partial x^2}(x, t) = (1 + a)\frac{\partial\theta}{\partial t}(x, t) - a\frac{d}{dt}\int_0^1 \theta(y, t)\, dy. \tag{2.1.3}$$

Equation (2.1.3) is an integro-differential equation in which the second term on the right-hand side is a function of t only, though not a function that is usually known to us in advance. If we were to ignore coupling and set $a = 0$, the equation would reduce to the familiar heat equation. The equation is intermediate in difficulty between the heat equation and the fourth-order equation (1.1.4), but its status differs from the status of each of those equations in that its form depends upon the conditions (1.1.3) which the displacement satisfies on the faces $x = 0$ and $x = 1$; any alteration in those conditions, such as demanding that the faces be stress-free, would generally lead to a different version of (2.1.3).

It should be noted that (2.1.2) implies that

$$\int_0^1 \left(\frac{\partial u}{\partial x}\right)^2 dx = a\int_0^1 \theta^2\, dx - a\left(\int_0^1 \theta\, dx\right)^2$$

and, therefore, that in the present context ($b = 0$) the total energy reduces to

$$E = \tfrac{1}{2}(1 + a)\int_0^1 \theta^2\, dx - \tfrac{1}{2}a\left(\int_0^1 \theta\, dx\right)^2, \tag{2.1.4}$$

and satisfies the inequalities

$$\tfrac{1}{2}\int_0^1 \theta^2\, dx \leqq E \leqq \tfrac{1}{2}(1 + a)\int_0^1 \theta^2\, dx. \tag{2.1.5}$$

§2.2. Construction of Solutions

We proceed to construct the solution θ of the integro-differential equation which is defined on $[0, 1] \times [0, \infty)$ and which corresponds to the decay of an initial distribution of temperature f when the faces of the body are maintained at the reference temperature. The construction depends upon separation of variables and generalized FOURIER Analysis.†

† The results of Sections 2.2–2.4 were first published in [18].

Theorem 2. *If f is C^2 in $[0, 1]$ and $f(0) = f(1) = 0$, there is exactly one solution θ of equation (2.1.3) which is continuous in $[0, 1] \times [0, \infty)$ and C^∞ in $[0, 1] \times (0, \infty)$ and which satisfies the boundary conditions*

$$\theta|_{x=0} = \theta|_{x=1} = 0,$$

and the initial condition

$$\theta|_{t=0} = f.$$

There is no difficulty whatsoever in verifying that there can be *at most one* solution θ which meets the hypotheses of Theorem 2. For Theorem 1 implies that the derivative of the total energy satisfies the equation

$$E' + \int_0^1 \left(\frac{\partial\theta}{\partial x}\right)^2 dx = 0$$

and, therefore, that the total energy cannot increase. In the same way, the total energy associated with the *difference* of two solutions cannot increase and, since it vanishes when $t = 0$, the total energy associated with the difference must vanish throughout the interval $[0, \infty)$. We conclude, with the help of the inequalities (2.1.5), that the difference of two solutions must vanish throughout the half-strip $[0, 1] \times [0, \infty)$.

In order to construct the solution it will be convenient to introduce the *inner product*

$$(g, h) = \int_0^1 gh\, dx - \frac{a}{1 + a} \int_0^1 g\, dx \cdot \int_0^1 h\, dx, \tag{2.2.1}$$

of functions $g, h, \ldots$ which are continuous on $[0, 1]$. SCHWARZ's inequality implies that

$$\frac{1}{1 + a} \int_0^1 g^2\, dx \leqq (g, g) \leqq \int_0^1 g^2\, dx$$

and, therefore, that the *norm*

$$\|g\| = \sqrt{(g, g)}$$

is equivalent to the norm in the LEBESGUE space $L^2[0, 1]$.

The integro-differential equation (2.1.3) has separable solutions

$$s_n(x) \exp\left(\frac{-\lambda_n^2 t}{1 + a}\right), \qquad n = 1, 2, 3, \ldots,$$

in which the functions s_n and the numbers λ_n satisfy the eigenvalue problem

$$\left.\begin{aligned} s_n''(x) &= -\lambda_n^2\left(s_n(x) - \frac{a}{1+a}\int_0^1 s_n(y)\,dy\right), \\ s_n(0) &= s_n(1) = 0, \end{aligned}\right\} \tag{2.2.2}$$

the primes denoting derivatives with respect to x.

It is easily checked that the eigenvalues λ_n, arranged in order of increasing magnitude, are the positive zeros of the function

$$\lambda \sin \lambda + 2a(1 - \cos \lambda).$$

Since the coupling constant a is a small positive number, λ_n is close to, and slightly greater than, $n\pi$ if n is odd, but $\lambda_n = n\pi$ *exactly* if n is even.

The corresponding eigenfunctions, which we suppose to be normalized by the requirement $\|s_n\| = 1$, are

$$s_n(x) = \frac{\sin \lambda_n x + \sin \lambda_n(1 - x) - \sin \lambda_n}{D_n}$$

if n is odd, the denominator being

$$D_n = \sqrt{(1 - \cos \lambda_n)\left(1 - \frac{1}{\lambda_n}\sin \lambda_n\right)}.$$

If n is even the eigenfunctions are

$$s_n(x) = \sqrt{2} \cdot \sin n\pi x,$$

just as they are for the heat equation, to which the integro-differential equation reduces when $a = 0$.

The eigenfunctions are mutually orthogonal with respect to our inner product, that is to say

$$(s_m, s_n) = 0 \quad \text{if } m \neq n.$$

Moreover, they can be given the usual variational characterizations: thus s_1 minimizes the functional

$$g \to \int_0^1 |g'|^2\,dx$$

among all functions g which are continuous and piecewise C^1 in $[0, 1]$ and satisfy

$$g(0) = g(1) = 0, \qquad \|g\| = 1.$$

Each subsequent eigenfunction s_n minimizes the same functional among all such g which meet the orthogonality requirements

$$(g, s_1) = \cdots = (g, s_{n-1}) = 0$$

in addition.

With the aid of considerations from the Calculus of Variations,† we shall be able to conclude that the initial temperature f can be developed in a series of eigenfunctions provided we can be assured that the eigenfunctions are uniformly bounded and that the series

$$\sum_{n=1}^{\infty} \frac{1}{\lambda_n^2}$$

converges.

The first is true because $\sin \lambda_n < 0$ and $\cos \lambda_n < 0$ when n is odd and, therefore, the denominator $D_n > 1$. Thus $|s_n| \leqq 3$ in $[0, 1]$ whether n is odd or even. The second is true because $\lambda_n \geqq n\pi$ for every n and, accordingly, we know that

$$f = \sum_{n=1}^{\infty} (f, s_n)s_n, \tag{2.2.3}$$

the series converging absolutely and uniformly in $[0, 1]$.

Next, we remark that the hypotheses on the initial temperature f ensure that the coefficients in the series development (2.2.3) satisfy the estimate

$$(f, s_n) = O\left(\frac{1}{n^2}\right) \quad \text{as } n \to \infty.$$

For, we have

$$\begin{aligned} (f, s_n) &= \int_0^1 f s_n \, dx - \frac{a}{1+a} \int_0^1 f \, dx \cdot \int_0^1 s_n \, dx \\ &= \int_0^1 f(x)\left(s_n(x) - \frac{a}{1+a} \int_0^1 s_n(y) \, dy \right) dx \\ &= -\frac{1}{\lambda_n^2} \int_0^1 f s_n'' \, dx \end{aligned}$$

† The most suitable reference is to the notes [23] taken by J. W. Green of the lectures of H. Lewy. In order to establish what we require it suffices to make only minor modifications to the arguments of Sections 5–10 of Chapter II of those notes.

and, when we integrate by parts twice, we deduce that

$$(f, s_n) = -\frac{1}{\lambda_n^2}\int_0^1 f'' s_n \, dx.$$

Thus SCHWARZ's inequality, the lower bound $\lambda_n \geqq n\pi$, and the inequality

$$\int_0^1 s_n^2 \, dx \leqq (1 + a)\|s_n\|^2 = 1 + a$$

combine to yield the estimates

$$|(f, s_n)| \leqq \frac{A}{n^2}, \tag{2.2.4}$$

where

$$A = \frac{1}{\pi^2}\sqrt{(1 + a)\int_0^1 (f'')^2 \, dx}.$$

Finally, we construct the solution θ by setting

$$\theta(x, t) = \sum_{n=1}^{\infty} (f, s_n) s_n(x) \exp\left(\frac{-\lambda_n^2 t}{1 + a}\right). \tag{2.2.5}$$

In view of the estimates (2.2.4), and the fact that the eigenfunctions are uniformly bounded, this series converges absolutely and uniformly in $[0, 1] \times [0, \infty)$. It defines a function θ which is continuous in $[0, 1] \times [0, \infty)$, and has continuous derivatives of all orders in $[0, 1] \times (0, \infty)$, and which satisfies the integro-differential equation (2.1.3) and the boundary conditions $\theta|_{x=0} = \theta|_{x=1} = 0$ and the initial condition $\theta|_{t=0} = f$.

§2.3. Failure of the Maximum Principle

Now that we have constructed the solution to the boundary and initial value problem of Theorem 2, we are in position to examine certain consequences of admitting thermomechanical coupling to the theory of heat conduction which are at variance with the conclusions of FOURIER's theory.

When the coupling constant $a = 0$ the series (2.2.5) reduces to the series

$$2 \sum_{n=1}^{\infty} \left(\int_0^1 f(y) \sin n\pi y \, dy \right) \sin n\pi x \cdot \exp(-n^2\pi^2 t), \tag{2.3.1}$$

whose sum is the solution of the heat equation

$$\frac{\partial^2 \theta}{\partial x^2} = \frac{\partial \theta}{\partial t}$$

satisfying the boundary conditions

$$\theta|_{x=0} = \theta|_{x=1} = 0$$

and the initial condition

$$\theta|_{t=0} = f.$$

The familiar Maximum Principle embodies a very important qualitative feature of the heat equation.† In the present context the principle has the implication that: *if* $a = 0$ *and if* $f > 0$ *in the open interval* $(0, 1)$ *then* $\theta \geqq 0$ *in the half-strip* $[0, 1] \times [0, \infty)$.

This property of the classical solution is not readily apparent from the series (2.3.1) which defines it, but it becomes so if we remember that we can represent the solution as the integral

$$\theta(x, t) = \int_0^1 G(x, y, t) f(y) \, dy$$

in which the kernel is

$$G(x, y, t) = 2 \sum_{n=1}^{\infty} \sin n\pi x \cdot \sin n\pi y \cdot \exp(-n^2\pi^2 t).$$

JACOBI's imaginary transformation in the theory of theta functions enables us to express the kernel in the alternative form

$$\frac{1}{2\sqrt{\pi t}} \sum_{n=-\infty}^{\infty} \left[\exp\left(- \frac{(x - y + 2n)^2}{4t} \right) - \exp\left(- \frac{(x + y + 2n)^2}{4t} \right) \right],$$

† PROTTER and WEINBERGER [31] attribute the Maximum Principle to E. E. LEVI [27, 28].

and it follows with the help of a further result of JACOBI's, namely the representation of a theta function as an infinite product,† that $G > 0$ in $(0, 1) \times (0, 1) \times (0, \infty)$. This fact makes it plain that $\theta \geqq 0$ in $[0, 1] \times [0, \infty)$ if $f > 0$ in $(0, 1)$.

On returning to the coupled and quasi-static theory, in which the coupling constant a is positive, we are led to ask if these qualitative features of the classical theory, namely the nonnegative properties of the temperature and of the associated kernel, are preserved. What we find is that they are *not* preserved: they are peculiar to the classical theory and any amount of coupling, no matter how small, is sufficient to destroy them.

We show first that it is possible to have $f > 0$ in $(0, 1)$ and yet have $\theta < 0$ throughout the body at all sufficiently large times t.

Theorem 3. *There is a function* f *such that* (i) $f > 0$ *in* $(0, 1)$, (ii) $f(0) = f(1) = 0$, *and* (iii) f *is* C^2 *in* $[0, 1]$, *but the corresponding* θ, *defined as in Theorem* 2, *is strictly negative throughout the open half-strip* $(0, 1) \times (t_0, \infty)$ *for some* $t_0 > 0$.

As we shall see, it suffices to take f to be a function whose graph has a high but thin positive peak, situated just inside one of the faces $x = 0$ or $x = 1$, and which is approximately a DIRAC delta distribution.

In order to construct f we shall need to demonstrate two properties of the eigenfunctions, namely that the first eigenfunction satisfies

$$s_1(x) \geqq \frac{4}{D_1} (\sin \tfrac{1}{2}\lambda_1)^3 x(1 - x), \tag{2.3.2}$$

while all subsequent eigenfunctions satisfy

$$|s_n| \leqq (n + 1)Bs_1, \tag{2.3.3}$$

where B is a constant depending upon the coupling constant a only.

The inequality (2.3.2) tells us that the first eigenfunction is strictly positive in $(0, 1)$, as we should expect to be the case. Its proof depends upon writing $s_1(x)$ as the product

$$\frac{4}{D_1} \sin \tfrac{1}{2}\lambda_1 \cdot \sin \tfrac{1}{2}\lambda_1 x \cdot \sin \tfrac{1}{2}\lambda_1 (1 - x).$$

---

† See, for example, WIDDER [32, Theorem 5.1].

Since the first eigenvalue lies in $0 < \lambda_1 < 2\pi$, $\sin \frac{1}{2}\lambda_1 x$ is a concave function of x in the closed unit interval $[0, 1]$ and, therefore, the second and third factors in the product satisfy

$$\sin \tfrac{1}{2}\lambda_1 x \geqq (\sin \tfrac{1}{2}\lambda_1)x,$$
$$\sin \tfrac{1}{2}\lambda_1(1 - x) \geqq (\sin \tfrac{1}{2}\lambda_1)(1 - x),$$

in that interval. Thus (2.3.2) is correct.

On turning to the remaining eigenfunctions, we note that, when n is odd, we can express $s_n(x)$ in the form

$$\frac{\lambda_n}{D_n}(1 - x)\int_0^x (\cos \lambda_n y - \cos \lambda_n(1 - y))\, dy$$
$$- \frac{\lambda_n}{D_n} x \int_x^1 (\cos \lambda_n y - \cos \lambda_n(1 - y))\, dy.$$

Since $D_n > 1$ and $\lambda_n < (n + 1)\pi$, we have

$$|s_n(x)| \leqq 4(n + 1)\pi x(1 - x) \leqq \frac{\pi D_1(n + 1)}{(\sin \frac{1}{2}\lambda_1)^3} s_1(x).$$

On the other hand, $s_n(x)$ equals

$$\sqrt{2}\cdot n\pi(1 - x)\int_0^x \cos n\pi y\, dy - \sqrt{2}\cdot n\pi x \int_x^1 \cos n\pi y\, dy,$$

when n is even and, therefore,

$$|s_n(x)| \leqq 2^{3/2} n\pi x(1 - x) \leqq \frac{\pi D_1 n}{\sqrt{2}(\sin \frac{1}{2}\lambda_1)^3} s_1(x).$$

Thus, every subsequent eigenfunction satisfies the required estimate (2.3.3), with

$$B = \frac{\pi D_1}{(\sin \frac{1}{2}\lambda_1)^3}.$$

When t is large it is the leading term in the series (2.2.5) for θ that dominates the remainder. If we appeal to the estimates (2.2.4) on the coefficients (f, s_n) we can estimate the sum of the remaining terms as

$$\left|\sum_{n=2}^{\infty} (f, s_n)s_n(x) \exp\left(\frac{-\lambda_n^2 t}{1+a}\right)\right|$$

$$\leqq \sum_{n=2}^{\infty} \frac{A}{n^2} \cdot B(n+1)s_1(x) \exp\left(\frac{-\lambda_n^2 t}{1+a}\right)$$

$$\leqq ABs_1(x) \exp\left(\frac{-\lambda_1^2 t}{1+a}\right) \sum_{n=2}^{\infty} \exp\left(\frac{(\lambda_1^2 - \lambda_n^2)t}{1+a}\right),$$

where

$$\sum_{n=2}^{\infty} \exp\left(\frac{(\lambda_1^2 - \lambda_n^2)t}{1+a}\right) \to 0 \quad \text{as } t \to \infty,$$

and, therefore,

$$\theta(x, t) = [(f, s_1) + o(1)]s_1(x) \exp\left(\frac{-\lambda_1^2 t}{1+a}\right), \tag{2.3.4}$$

the term $o(1)$ being uniform with respect to x in $[0, 1]$.

Our assertion that f can be chosen in such a way that $f > 0$ in $(0, 1)$ but $\theta < 0$ in some half-strip $(0, 1) \times (t_0, \infty)$ rests upon the (somewhat unexpected) observation that even though s_1 is positive in $(0, 1)$ we can choose f in such a way that it too is positive in $(0, 1)$ but the inner product (f, s_1) is nonetheless negative.

For, the eigenfunction s_1 vanishes at $x = 0$ but $\sin \lambda_1 \neq 0$ if $a > 0$ and, therefore, there is a number ξ in $(0, 1)$ such that

$$s_1(\xi) < \frac{|\sin \lambda_1|}{D_1}.$$

Having chosen ξ, we note that arguments of a kind familiar from the theory of generalized functions assure us that there is a sequence of functions $\{f_n\}_{n \geqq 1}$ with the properties:

(1) $f_n > 0$ in $(0, 1)$,
(2) $f_n(0) = f_n(1) = 0$,
(3) f_n is C^∞ in $[0, 1]$,
(4) $\int_0^1 f_n \, dx = 1$,
(5) $\int_0^1 f_n s_1 \, dx \to s_1(\xi)$ as $n \to \infty$.

Now consider the inner product (f_n, s_1) for arbitrary n. It is

$$(f_n, s_1) = \int_0^1 f_n s_1 \, dx - \frac{a}{1+a} \int_0^1 f_n \, dx \cdot \int_0^1 s_1 \, dx$$
$$= \int_0^1 f_n s_1 \, dx - \frac{a}{1+a} \int_0^1 s_1 \, dx,$$

where, as a straightforward calculation using the known form of s_1, shows

$$\frac{a}{1+a} \int_0^1 s_1 \, dx = \frac{|\sin \lambda_1|}{D_1}.$$

It follows that

$$(f_n, s_1) \to s_1(\xi) - \frac{|\sin \lambda_1|}{D_1} < 0 \quad \text{as } n \to \infty,$$

and, therefore, if we take $f = f_n$ for some sufficiently large n we shall have $(f, s_1) < 0$ and, at the same time, $f > 0$ in $(0, 1)$, $f(0) = f(1) = 0$, and f will be C^∞ in $[0, 1]$.

With f chosen in this way, (2.3.4) shows that θ is strictly negative throughout $(0, 1) \times (t_0, \infty)$ provided that t_0 is sufficiently large, which is what is required.

§2.4. Behaviour of the Kernel

The solution of the boundary and initial value problem constructed in Theorem 2 can be represented as the integral

$$\theta(x, t) = \int_0^1 K(x, y, t) f(y) \, dy \tag{2.4.1}$$

in which the kernel $K(x, y, t)$ equals

$$\sum_{n=1}^{\infty} s_n(x)\left(s_n(y) - \frac{a}{1+a} \int_0^1 s_n(z) \, dz\right) \exp\left(\frac{-\lambda_n^2 t}{1+a}\right).$$

Since, as we can readily verify,

$$\frac{a}{1+a} \int_0^1 s_n(z) \, dz = -\frac{\sin \lambda_n}{D_n}$$

if n is odd, and

$$\frac{a}{1+a} \int_0^1 s_n(z) \, dz = 0$$

if n is even, the kernel equals

$$\sum_{n\,\text{odd}} \frac{1}{D_n^2}(\sin \lambda_n x + \sin \lambda_n(1-x) - \sin \lambda_n)(\sin \lambda_n y + \sin \lambda_n(-y))$$

$$\cdot \exp\left(\frac{-\lambda_n^2 t}{1+a}\right) + 2 \sum_{n\,\text{even}} \sin n\pi x \cdot \sin n\pi y \cdot \exp\left(\frac{-n^2\pi^2 t}{1+a}\right),$$

where the first sum is taken over all odd positive integers and the second over all even positive integers. The series converge absolutely and uniformly in $[0, 1] \times [0, 1] \times [\delta, \infty)$ for each $\delta > 0$, and define a kernel K which is continuous in $[0, 1] \times [0, 1] \times (0, \infty)$.

As we should expect, K reduces, when $a = 0$, to the kernel G associated with the heat equation.

By contrast with the classical case, Theorem 3 shows that, when $a > 0$, the kernel $K(x, y, t)$ must be *negative* at some point (x, y) of the unit square $[0, 1] \times [0, 1]$, at least if t is sufficiently large. In fact, the kernel must have changes of sign in the unit square for each positive t, whether large or not; the proof that this is so involves considering the series of positive terms

$$P(t) = 4(1+a) \sum_{n\,\text{odd}} \left(\frac{1 - \cos \lambda_n}{D_n \lambda_n}\right)^2 \exp\left(\frac{-\lambda_n^2 t}{1+a}\right),$$

which converges at every t in $[0, \infty)$.

Theorem 4. *Let $t > 0$ be arbitrary. Then there are points (x, y) in $[0, 1] \times [0, 1]$ at which $K(x, y, t) \geqq P(t)$, and there are points at which $K(x, y, t) \leqq -aP(t)$.*

The subset of the unit square in which the kernel is negative must be expected to vary with t.

The assertions of Theorem 4 are consequences of the equations

$$\int_0^1 \int_0^1 K(x, y, t)\, dx\, dy = \sum_{n\,\text{odd}} \frac{1}{D_n^2} \left(\frac{2}{\lambda_n}(1 - \cos \lambda_n) - \sin \lambda_n\right)$$

$$\cdot \frac{2}{\lambda_n}(1 - \cos \lambda_n) \cdot \exp\left(\frac{-\lambda_n^2 t}{1+a}\right),$$

and

$$\int_0^1 K(x, 0, t)\, dx = \sum_{n\,\text{odd}} \frac{1}{D_n^2} \left(\frac{2}{\lambda_n}(1 - \cos \lambda_n) - \sin \lambda_n\right) \cdot \sin \lambda_n \cdot \exp\left(\frac{-\lambda_n^2 t}{1+a}\right),$$

for, if we recall that λ_n is a zero of the function

$$\lambda \sin \lambda + 2a(1 - \cos \lambda),$$

we can replace $\sin \lambda_n$ by $-2a(1 - \cos \lambda_n)/\lambda_n$ to arrive at the formulae

$$\int_0^1 \int_0^1 K(x, y, t)\, dx\, dy = P(t),$$

$$\int_0^1 K(x, 0, t)\, dx = -aP(t), \tag{2.4.2}$$

and, thus, the theorem is correct.

Equation (2.4.2) can also be used to show that it is not in fact necessary to wait a long time in order to violate the conclusion of the Maximum Principle. Indeed, given any $t > 0$, no matter how small, we can choose f in such a way that $f > 0$ in $(0, 1)$ but the mean temperature

$$\int_0^1 \theta(x, t)\, dx$$

at the given time is negative. The choice of f will depend upon the assigned value of t. The reader is referred to [18] for the proof of this result.

§2.5. Initial Sensitivity to the Boundary†

In this section we continue our investigation into the behaviour of the solution θ to the boundary and initial value problem considered in Theorem 2, and once again we arrive at conclusions which, because we have admitted coupling to the theory of heat conduction, are at variance with the conclusions of FOURIER's theory. Our considerations shed further light on the behaviour of the kernel K which occurs in the integral representation for θ, and on the manner in which K differs from G, the corresponding kernel for the heat equation.

---

† Corresponding results for the theory of heat conduction in a thermoelastic fluid, and in dimension 3, were published in [19].

The kernels $K(x, y, t)$ and $G(x, y, t)$ are, of course, *fundamental solutions*, that is to say they are the solutions, of the integro-differential equation (2.1.3) and of the heat equation, respectively, which vanish when $x = 0$ and $x = 1$ and correspond initially to a delta distribution of temperature concentrated at the point y.

As $t \to 0+$, the classical kernel has the asymptotic behaviour

$$G(x, y, t) \sim \frac{1}{2\sqrt{\pi t}} \exp\left(\frac{-(x-y)^2}{4t}\right) \tag{2.5.1}$$

at points x, y of the open interval $(0, 1)$. The function to which G is asymptotic is just the *source solution*, that is to say the fundamental solution which results if we replace the bounded interval $[0, 1]$ by the real line $(-\infty, \infty)$. In this sense the solution is initially insensitive, at interior points, to the presence of the boundary points $x = 0$ and $x = 1$ at which the temperature is required to vanish.†

If we ask whether the same conclusion is valid for the integro-differential equation (2.1.3), the answer is less accessible to intuition in view of the fact that the interval $[0, 1]$ enters the problem not just by way of the boundary conditions and the initial condition but figures directly in the integro-differential equation as a domain of integration.

For our purpose it will be convenient to recast (2.5.1) as a statement about the asymptotic behaviour, as $s \to \infty$, of the LAPLACE transform

$$\hat{G}(x, y, s) = \int_0^\infty \exp(-st) G(x, y, t)\, dt.$$

As is well known, this transform equals

$$\begin{cases} \dfrac{\sinh\sqrt{s}x \cdot \sinh\sqrt{s}(1-y)}{\sqrt{s}\sinh\sqrt{s}}, & 0 \leqq x \leqq y \leqq 1, \\[2ex] \dfrac{\sinh\sqrt{s}(1-x) \cdot \sinh\sqrt{s}y}{\sqrt{s}\sinh\sqrt{s}}, & 0 \leqq y \leqq x \leqq 1, \end{cases} \tag{2.5.2}$$

† KAC [25, 26] has made the corresponding observation for the heat equation in higher dimensions the basis from which to deduce certain very interesting results concerning the eigenvalues of the LAPLACE operator. In KAC's vivid language, the solution does not *feel* the boundary initially.

and its asymptotic behaviour is

$$\hat{G}(x, y, s) \sim \frac{1}{2\sqrt{s}} \exp(-\sqrt{s}|x - y|) \quad \text{as } s \to \infty, \tag{2.5.3}$$

whenever x and y are points of the open interval $(0, 1)$. The function to which $\hat{G}$ is asymptotic is, of course, just the transform of the source solution.

In order to state our next result, which concerns the asymptotic behaviour of the transform

$$\hat{K}(x, y, s) = \int_0^\infty \exp(-st)K(x, y, t)\, dt,$$

it will be convenient to write

$$d(y) = \min(y, 1 - y)$$

for the distance from y to the nearer of the end points of the interval $[0, 1]$, and to introduce the sets

$$I(y) = \{x: 0 < x < 1 \text{ and } |x - y| < d(y)\},$$
$$J(y) = \{x: 0 < x < 1 \text{ and } |x - y| > d(y)\}.$$

If $0 < y < \frac{1}{2}$, $I(y)$ is the interval $(0, 2y)$ and $J(y)$ is the interval $(2y, 1)$; if $\frac{1}{2} < y < 1$, $I(y)$ is the interval $(2y - 1, 1)$ and $J(y)$ is the interval $(0, 2y - 1)$; if $y = \frac{1}{2}$, $I(y)$ is the entire interval $(0, 1)$ and $J(y)$ is empty.

Theorem 5. *Let y be any point of $(0, 1)$. Then, as $s \to \infty$,*

$$\hat{K}(x, y, s) \sim \frac{1}{2}\sqrt{\frac{1 + a}{s}} \exp(-\sqrt{(1 + a)s} \cdot |x - y|) \tag{2.5.4}$$

if x is a point of $I(y)$, but

$$\hat{K}(x, y, s) \sim -\frac{a}{s} \exp(-\sqrt{(1 + a)s} \cdot d(y)) \tag{2.5.5}$$

if x is a point of $J(y)$.

Before proceeding to the proof of Theorem 5 we make a number of comments upon it.

It will be seen, first of all, that the right-hand side of (2.5.4) is none other than the LAPLACE transform of the source solution associated with the heat equation

$$\frac{\partial^2 \theta}{\partial x^2} = (1 + a)\frac{\partial \theta}{\partial t},$$

in which the (scaled) diffusivity has the value $(1 + a)^{-1}$. Since this transform depends upon the points x and y only through the distance between them, and not upon the position of y relative to the end points of the interval $[0, 1]$, we may say that:

Each point x of $I(y)$ is a point of initial insensitivity to the boundary, with respect to y.

Next, we note that the right-hand side of (2.5.5) does depend upon the position of y relative to the end points and, hence, we may say, by constrast with the classical behaviour, that:

Each point x of $J(y)$ is a point of initial sensitivity to the boundary, with respect to y.

Furthermore, the interval $J(y)$ is determined solely by the position of y in $(0, 1)$ and is quite independent of the size of the coupling constant a so long as a is positive. The significance of this fact is that, if a is small, we should expect the presence of the terms

$$a\frac{\partial \theta}{\partial t}(x, t) - a\frac{d}{dt}\int_0^1 \theta(y, t)\, dy$$

in the integro-differential equation to produce only small *quantitative* changes in the temperature, but Theorem 5 reinforces the conclusion at which we arrived in Sections 2.3 and 2.4, namely that any amount of coupling, no matter how small, alters the *qualitative* behaviour.

Finally, we note that the right-hand side of (2.5.4) is negative, in agreement with our conclusion that the kernel K must change sign. In fact, we can now say a little more.

Theorem 6. *Let y be any point of $(0, 1)$ other than $y = \frac{1}{2}$, and let x be any point of $J(y)$. Then there are values of $t > 0$, depending upon x and y, such that*

$$K(x, y, t) < 0.$$

Our proof of Theorem 5 involves constructing the transform $\hat{K}$ in terms of $\hat{G}$. We shall need the fact that the latter transform, whose values are listed in (2.5.2), is none other than the GREEN's function which occurs in the integral representation

$$w(x) = \int_0^1 \hat{G}(x, y, s)g(y)\,dy$$

for the solution w of the boundary value problem

$$\frac{d^2w}{dx^2} = sw - g, \qquad w|_{x=0} = w|_{x=1} = 0.$$

Now let us return to the boundary and initial value problem for the integro-differential equation that we considered in Theorem 2. The transform $\hat{\theta}$ is determined by solving the auxiliary boundary value problem

$$\frac{d^2\hat{\theta}}{dx^2}(x, s) = (1 + a)(s\hat{\theta}(x, s) - f(x)) - a\int_0^1 (s\hat{\theta}(y, s) - f(y))\,dy,$$

$$\hat{\theta}|_{x=0} = \hat{\theta}|_{x=1} = 0.$$

In view of the interpretation of $\hat{G}$ as a GREEN's function, it must be that $\hat{\theta}$ satisfies the FREDHOLM integral equation

$$\hat{\theta}(x, s) = \int_0^1 \hat{G}(x, y, (1 + a)s)\left[(1 + a)f(y) + a\int_0^1 (s\hat{\theta}(z, s) - f(z))\,dz\right]dy,$$

that is to say

$$\begin{aligned}\hat{\theta}(x, s) = \int_0^1 \bigg[&(1 + a)\hat{G}(x, y, (1 + a)s) \\ &- a\int_0^1 \hat{G}(x, z, (1 + a)s)\,dz\bigg]f(y)\,dy \\ &+ as\int_0^1 \hat{G}(x, y, (1 + a)s)\,dy \cdot \int_0^1 \hat{\theta}(z, s)\,dz.\end{aligned}$$

On integrating both sides of this equation with respect to x we can deduce the value of the integral

$$\int_0^1 \hat{\theta}(z, s)\,dz$$

and, thus, we can determine the transform $\hat{\theta}$ itself. If we carry through the details of the calculation, which are straightforward, and remember that the transform must equal

$$\hat{\theta}(x, s) = \int_0^1 \hat{K}(x, y, s) f(y)\, dy,$$

we can then express $\hat{K}$ in terms of $\hat{G}$. It transpires that the difference

$$\hat{K}(x, y, s) - (1 + a)\hat{G}(x, y, (1 + a)s)$$
$$= \frac{a \displaystyle\int_0^1 \hat{G}(x, z^*, (1 + a)s)\, dz^* \left[(1 + a)s \int_0^1 \hat{G}(z, y, (1 + a)s)\, dz - 1 \right]}{1 - as \displaystyle\int_0^1 \int_0^1 \hat{G}(z, z^*, (1 + a)s)\, dz\, dz^*}.$$

Since we know $\hat{G}$ explicitly we can determine the right-hand side and, indeed, we find, after further calculation, that

$$\hat{K}(x, y, s) - (1 + a)\hat{G}(x, y, (1 + a)s) = \frac{-\dfrac{a}{s}(1 - F(x, s, a))F(y, s, a)}{1 + \dfrac{2a}{\sqrt{(1 + a)s}} \tanh \frac{1}{2}\sqrt{(1 + a)s}},$$

where

$$F(y, s, a) = \frac{\sinh(\sqrt{(1 + a)s} \cdot y) + \sinh(\sqrt{(1 + a)s} \cdot (1 - y))}{\sinh \sqrt{(1 + a)s}}.$$

The asymptotic behaviour of F is

$$F(y, s, a) \sim \exp(-\sqrt{(1 + a)s} \cdot d(y)) \quad \text{as } s \to \infty,$$

where $d(y)$ is the distance introduced earlier, and so we conclude that

$$\hat{K}(x, y, s) - (1 + a)\hat{G}(x, y, (1 + a)s)$$
$$\sim -\frac{a}{s} \exp(-\sqrt{(1 + a)s} \cdot d(y)) \quad \text{as } s \to \infty. \qquad (2.5.6)$$

Since, as we know

$$(1 + a)\hat{G}(x, y, (1 + a)s) \sim \frac{1}{2}\sqrt{\frac{1 + a}{s}} \cdot \exp(-\sqrt{(1 + a)s} \cdot |x - y|)$$

it is clear from (2.5.6) that the asymptotic behaviour of $\hat{K}$ is that indicated by (2.5.4) if $|x - y| < d(y)$, that is if x belongs to $I(y)$, but the asymptotic behaviour is that indicated by (2.5.5) if $|x - y| > d(y)$, that is if x belongs to $J(y)$.

§2.6. A Monotone Property of the Entropy

Thus far, in the course of examining the behaviour of the solution of the boundary and initial value problem of Theorem 2, we have pointed out features which, because of the presence of coupling, are qualitatively different from those of the corresponding solution for the heat equation.

We turn now to a qualitative property which is similar to a known property of the heat equation, but even here there is a difference in that what is true of the temperature according to FOURIER's theory becomes a property of the *entropy* according to the coupled and quasi-static theory of thermoelasticity.

It will be recalled that the entropy is

$$\eta = \theta + \sqrt{a} \cdot \frac{\partial u}{\partial x}$$

and that the strain is given by equation (2.1.2). Thus the entropy is

$$\eta(x, t) = (1 + a)\theta(x, t) - a \int_0^1 \theta(y, t)\, dy. \tag{2.6.1}$$

We begin with the following remark;

Theorem 7. *The hypotheses of Theorem 2 imply that*

$$E, \quad \int_0^1 \theta^2\, dx, \quad \int_0^1 \eta^2\, dx$$

all decay to zero at an exponential rate as $t \to \infty$.

We can either verify this directly from the series expansion (2.2.5) or, alternatively, by means of an energy integral argument. Thus, to take the second course, we start from the identity

$$E' + \int_0^1 \left(\frac{\partial \theta}{\partial x}\right)^2 dx = 0$$

and note that,† because θ vanishes at $x = 0$ and $x = 1$,

$$\int_0^1 \left(\frac{\partial\theta}{\partial x}\right)^2 dx \geqq \pi^2 \int_0^1 \theta^2\, dx.$$

Since, as we know,

$$\int_0^1 \theta^2\, dx \geqq \frac{2}{1+a} E$$

we conclude that the total energy satisfies the differential inequality

$$E' + \frac{2\pi^2}{1+a} E \leqq 0$$

and, therefore, that

$$0 \leqq E(t) \leqq E(0) \exp\left(\frac{-2\pi^2 t}{1+a}\right).$$

Thus E decays to zero at an exponential rate.

The inequalities

$$\int_0^1 \theta^2\, dx \leqq 2E,$$

$$\int_0^1 \eta^2\, dx = (1+a)^2 \int_0^1 \theta^2\, dx - (2a + a^2)\left(\int_0^1 \theta\, dx\right)^2 \leqq 2(1+a)^2 E,$$

enable us to conclude that the same is true of the integrals

$$\int_0^1 \theta^2\, dx, \quad \int_0^1 \eta^2\, dx.$$

There is more to be said, though, about the manner in which the entropy decays to zero and we shall, in fact, establish a *monotone* relationship between the entropy and the time t.

Monotone relationships between the entropy and the time are of especial interest in Thermodynamics but such relationships usually involve the total entropy of a body; the result proved here is of a different type in that it refers to a maximum value taken with respect to position in the body.

We consider once again the boundary and initial value problem of Theorem 2. We impose no restriction upon the sign of the initial

† Hardy, Littlewood, and Pólya [24, Theorem 257].

temperature f, but we do find it necessary to restrict the size of the coupling constant a; the restriction, though, is a physically realistic one.

Theorem 8. *If* $0 \leqq a \leqq 1$, *the hypotheses of Theorem* 2 *imply that*

$$\max\{|\eta(x, t)|: 0 \leqq x \leqq 1\}$$

is a decreasing function of t *in* $[0, \infty)$.

When $a = 0$, Theorem 8 reduces to a known result for the heat equation which is due to PÓLYA and SZEGÖ [30] and asserts that $\max|\theta|$ is a decreasing function. BELLMAN [1] gave a different proof of PÓLYA and SZEGÖ's result, and our own proof of Theorem 8 involves making suitable modifications to BELLMAN's argument.

In order to prove Theorem 8 we observe that we can write the integro-differential equation (2.1.3) as

$$\frac{\partial^2 \theta}{\partial x^2} = \frac{\partial \eta}{\partial t},$$

and that the second derivative of the entropy with respect to x is

$$\frac{\partial^2 \eta}{\partial x^2} = (1 + a)\frac{\partial^2 \theta}{\partial x^2}.$$

Thus, the entropy is a solution of the heat equation

$$\frac{\partial^2 \eta}{\partial x^2} = (1 + a)\frac{\partial \eta}{\partial t}.$$

It is, however, a solution which satisfies boundary conditions of a somewhat unusual kind, for the equation (2.6.1) which expresses η in terms of θ implies that

$$\int_0^1 \eta\, dx = \int_0^1 \theta\, dx.$$

Accordingly,

$$(1 + a)\theta(x, t) = \eta(x, t) + a\int_0^1 \eta(y, t)\, dy$$

and, since $\theta|_{x=0} = \theta|_{x=1} = 0$ it must be that η satisfies the nonlocal boundary conditions†

$$\eta(0, t) = \eta(1, t) = -a \int_0^1 \eta(x, t)\, dx, \qquad t \geqq 0, \tag{2.6.2}$$

which connect the entropy at the boundary with the total entropy of the body.

Guided by BELLMAN's argument,‡ we consider the functions

$$I_n = \int_0^1 \eta^{2n}\, dx + a^{2n-1}\left(\int_0^1 \eta\, dx\right)^{2n},$$

n being any positive integer. The derivative of I_n is

$$I_n' = 2n \int_0^1 \eta^{2n-1} \frac{\partial \eta}{\partial t}\, dx + 2na^{2n-1}\left(\int_0^1 \eta\, dx\right)^{2n-1} \cdot \frac{d}{dt} \int_0^1 \eta\, dx.$$

If we appeal to the heat equation that η satisfies and use integration by parts we can rewrite the integral

$$\int_0^1 \eta^{2n-1} \frac{\partial \eta}{\partial t}\, dx$$

as

$$\frac{1}{1+a} \int_0^1 \eta^{2n-1} \frac{\partial^2 \eta}{\partial x^2}\, dx = \frac{1}{1+a}\left(\eta^{2n-1} \frac{\partial \eta}{\partial x}\right)\bigg|_{x=1} - \frac{1}{1+a}\left(\eta^{2n-1} \frac{\partial \eta}{\partial x}\right)\bigg|_{x=0}$$
$$- \frac{(2n-1)}{1+a} \int_0^1 \eta^{2n-2}\left(\frac{\partial \eta}{\partial x}\right)^2 dx,$$

or, with the aid of the boundary conditions, as

$$-\frac{a^{2n-1}}{1+a}\left(\int_0^1 \eta\, dx\right)^{2n-1}\left(\frac{\partial \eta}{\partial x}\bigg|_{x=1} - \frac{\partial \eta}{\partial x}\bigg|_{x=0}\right)$$
$$- \frac{(2n-1)}{1+a} \int_0^1 \eta^{2n-2}\left(\frac{\partial \eta}{\partial x}\right)^2 dx.$$

† See [12, 13]. My results have been extensively generalized by FRIEDMAN [22] to embrace parabolic equations in $\mathbb{R}^n$ which satisfy nonlocal boundary conditions.

‡ In the classical case $a = 0$ and the second term in the definition of I_n is absent from BELLMAN's argument; likewise the troublesome integral is then absent from the boundary conditions (2.6.2).

Since, as integration of the heat equation with respect to x shows,

$$\left.\frac{\partial\eta}{\partial x}\right|_{x=1} - \left.\frac{\partial\eta}{\partial x}\right|_{x=0} = \int_0^1 \frac{\partial^2\eta}{\partial x^2}\,dx = (1+a)\frac{d}{dt}\int_0^1 \eta\,dx,$$

the integral

$$\int_0^1 \eta^{2n-1}\frac{\partial\eta}{\partial t}\,dx = -a^{2n-1}\left(\int_0^1 \eta\,dx\right)^{2n-1}\cdot\frac{d}{dt}\int_0^1 \eta\,dx$$
$$-\frac{(2n-1)}{1+a}\int_0^1 \eta^{2n-2}\left(\frac{\partial\eta}{\partial x}\right)^2 dx$$

and, therefore, the derivative

$$I_n' = -\frac{2n(2n-1)}{1+a}\int_0^1 \eta^{2n-2}\left(\frac{\partial\eta}{\partial x}\right)^2 dx \leqq 0.$$

Thus, each I_n is a decreasing function, and we shall have completed the proof once we have shown that

$$I_n^{1/2n} \to \max|\eta| \quad \text{as } n \to \infty.$$

To make this last step we appeal to the restriction $0 \leqq a \leqq 1$, and to the inequality

$$\left(\int_0^1 \eta\,dx\right)^{2n} \leqq \int_0^1 \eta^{2n}\,dx,$$

which is an instance of HÖLDER's inequality (and also of JENSEN's inequality for convex functions), to deduce that

$$\int_0^1 \eta^{2n}\,dx \leqq I_n \leqq (1+a^{2n-1})\int_0^1 \eta^{2n}\,dx \leqq 2\int_0^1 \eta^{2n}\,dx.$$

Since

$$\left(\int_0^1 \eta^{2n}\,dx\right)^{1/2n} \to \max|\eta| \quad \text{and} \quad 2^{1/2n} \to 1 \quad \text{as } n \to \infty,$$

the proof is complete.

CHAPTER 3

Trigonometric Solutions of the Integro-differential Equation

§3.1. Maximum Principles for the Pointwise Mean Total Energy Density and the Pointwise Mean Square Heat Flux

In this chapter we examine further the consequences† of the coupled and quasi-static approximation which leads to the integro-differential equation (2.1.3) for the temperature. We now abandon the boundary and initial value problem which was formulated in Theorem 2 and proceed to study a *trigonometric solution* of the integro-differential equation, that is to say a solution which has the form

$$\theta(x, t) = \operatorname{Re} \sum \Theta(x, \omega) \exp(i\omega t), \tag{3.1.1}$$

where Θ is a (complex-valued) solution of the integro-differential equation

$$\Theta''(x, \omega) = i\omega\left[(1 + a)\Theta(x, \omega) - a \int_0^1 \Theta(y, \omega)\, dy\right], \tag{3.1.2}$$

in which the primes denote derivatives with respect to x, and the sum (3.1.1) is taken over a finite set of distinct positive real exponents ω.

The sum (3.1.1) is not usually a periodic function of t, but it is always a uniformly almost periodic function, and we shall always

† The results of this chapter are to be found, for the most part, in [16].

understand the *mean value operator* M, taken with respect to the time t, to be defined as in the Theory of Almost Periodic Functions or as in WIENER's Generalized Harmonic Analysis,† that is to say

$$M\{\cdots\} = \lim_{T\to\infty} \frac{1}{2T}\int_{-T}^{T} \cdots\, dt.$$

We begin by establishing that the *pointwise mean values*

$$M\{e, x\} = \lim_{T\to\infty} \frac{1}{2T}\int_{-T}^{T} e(x, t)\, dt,$$

$$M\{q^2, x\} = \lim_{T\to\infty} \frac{1}{2T}\int_{-T}^{T} q(x, t)^2\, dt,$$

of the total energy density e and of the square of the heat flux q, always assume their maximum values at $x = 0$ or at $x = 1$.

When the coupling constant $a = 0$, e reduces to $\frac{1}{2}\theta^2$ and, thus, within the classical theory, the pointwise mean square temperature, associated with a trigonometric solution, assumes its maximum value at $x = 0$ or at $x = 1$; it is a simple matter to prove this property of the heat equation directly but I am not aware that attention has been drawn to it hitherto.

Theorem 9. *If θ is a trigonometric solution of the integro-differential equation* (2.1.3), *the mean values $M\{e, x\}$ and $M\{q^2, x\}$ exist at each point x of the interval* $[0, 1]$. *Moreover, each is a convex function of x and, in particular, each assumes its maximum value with respect to* $[0, 1]$ *at $x = 0$, or at $x = 1$.*

It will be recalled that in the present context, in which the inertial constant $b = 0$, the total energy density is

$$e = \frac{1}{2}\left(\theta^2 + \left(\frac{\partial u}{\partial x}\right)^2\right),$$

that is to say

$$e(x, t) = \tfrac{1}{2}\theta(x, t)^2 + \tfrac{1}{2}a\left(\theta(x, t) - \int_0^1 \theta(y, t)\, dy\right)^2.$$

† See BESICOVITCH [2] and WIENER [33, 34]. We require no knowledge of those theories other than the capacity to calculate the result of applying the mean value operator to finite trigonometric sums. In my view, WIENER's abundantly fertile ideas in the area of Generalized Harmonic Analysis have not received the attention they deserve from students of Continuum Mechanics.

Since θ is the finite trigonometric sum (3.1.1), standard arguments of the Theory of Almost Periodic Functions assure us that the mean value $M\{e, x\}$ exists at each x in $[0, 1]$ and equals

$$\tfrac{1}{4}\sum |\Theta(x, \omega)|^2 + \tfrac{1}{4}a \sum \left|\Theta(x, \omega) - \int_0^1 \Theta(y, \omega)\, dy\right|^2.$$

Similarly, the mean value $M\{q^2, x\}$ of the square of the heat flux $q = -\partial\theta/\partial x$ exists at each x in $[0, 1]$ and equals

$$\tfrac{1}{2}\sum |\Theta'(x, \omega)|^2.$$

Thus it will be enough to prove that each of

$$|\Theta(x, \omega)|^2 + a\left|\Theta(x, \omega) - \int_0^1 \Theta(y, \omega)\, dy\right|^2, \tag{3.1.3}$$

$$|\Theta'(x, \omega)|^2, \tag{3.1.4}$$

is a convex function of x, and to do so it is enough to examine their second derivatives with respect to x. In fact, if we denote complex conjugates by a bar we find the second derivative of (3.1.3) to be

$$\begin{aligned}
&\frac{d^2}{dx^2}\left\{\Theta(x, \omega)\overline{\Theta(x, \omega)} + a\left(\Theta(x, \omega) - \int_0^1 \Theta(y, \omega)\, dy\right)\right.\\
&\qquad\qquad\left.\cdot\left(\overline{\Theta(x, \omega)} - \int_0^1 \overline{\Theta(y, \omega)\, dy}\right)\right\}\\
&= \Theta''(x, \omega)\overline{\Theta(x, \omega)} + 2\Theta'(x, \omega)\overline{\Theta'(x, \Theta)} + \Theta(x, \omega)\overline{\Theta''(x, \omega)}\\
&\quad + a\Theta''(x, \omega)\left(\overline{\Theta(x, \omega)} - \int_0^1 \overline{\Theta(y, \omega)}\, dy\right)\\
&\quad + 2a\Theta'(x, \omega)\overline{\Theta'(x, \omega)}\\
&\quad + a\left(\Theta(x, \omega) - \int_0^1 \Theta(y, \omega)\, dy\right)\overline{\Theta''(x, \omega)}\\
&= \Theta''(x, \omega)\left[(1 + a)\overline{\Theta(x, \omega)} - a\int_0^1 \overline{\Theta(y, \omega)}\, dy\right]\\
&\quad + 2(1 + a)|\Theta'(x, \omega)|^2\\
&\quad + \overline{\Theta''(x, \omega)}\left[(1 + a)\Theta(x, \omega) - a\int_0^1 \Theta(y, \omega)\, dy\right].
\end{aligned}$$

On substituting for $\Theta''(x, \omega)$ from (3.1.2), and for $\overline{\Theta''(x, \omega)}$ from the conjugate equation, namely

$$\overline{\Theta''(x, \omega)} = -i\omega\left[(1 + a)\overline{\Theta(x, \omega)} - a \int_0^1 \overline{\Theta(y, \omega)}\, dy\right],$$

we conclude that the second derivative of (3.1.3) equals

$$2(1 + a)|\Theta'(x, \omega)|^2$$

and is, therefore, nonnegative. Thus, (3.1.3) is indeed a convex function.

A similar argument establishes that the second derivative of the function (3.1.4) is

$$2|\Theta''(x, \omega)|^2 \geqq 0$$

and so (3.1.4) also is convex and the proof is complete.

§3.2. The Effect of Coupling on Trigonometric Solutions

If θ is a trigonometric solution of the integro-differential equation (2.1.3), the *boundary temperatures*

$$f = \theta|_{x=0}, \qquad g = \theta|_{x=1}, \tag{3.2.1}$$

must themselves be real trigonometric sums of the forms

$$\left.\begin{aligned} f(t) &= \operatorname{Re} \sum F(\omega) \exp(i\omega t), \\ g(t) &= \operatorname{Re} \sum G(\omega) \exp(i\omega t), \end{aligned}\right\} \tag{3.2.2}$$

in which the coefficients $F(\omega)$ and $G(\omega)$ are complex.

In what follows we shall think of f and g as being prescribed independently of the value of the coupling constant a and we shall ask how θ depends upon a; we are, in effect, comparing the steady state oscillatory responses, to prescribed oscillatory boundary temperatures, of different bodies which have the same thicknesses but which have different coupling constants.

It is a straightforward matter to construct the (unique) trigonometric solution which satisfies (3.2.1). For, if we put

$$\phi(x, \omega, a) = \frac{\sinh px + \dfrac{a(\cosh p - 1)}{p \sinh p}(\sinh px - \sinh p(1 - x) + \sinh p)}{\sinh p + \dfrac{2a}{p}(\cosh p - 1)}, \tag{3.2.3}$$

where

$$p(\omega, a) = (1 + i)\sqrt{(1 + a)\omega/2}$$

then ϕ satisfies the integro-differential equation

$$\phi''(x, \omega, a) = i(1 + a)\omega\phi(x, \omega, a) - ia\omega \int_0^1 \phi(y, \omega, a)\, dy$$

and the boundary conditions

$$\phi(0, \omega, a) = 0, \qquad \phi(1, \omega, a) = 1,$$

and, therefore, the sum (3.1.1) is the desired trigonometric solution provided that $\Theta(x, \omega)$ is defined to be

$$F(\omega)\phi(1 - x, \omega, a) + G(\omega)\phi(x, \omega, a).$$

The dependence of θ upon the coupling constant at individual points (x, t) of $[0, 1] \times (-\infty, \infty)$ appears to be too complicated to be of interest, and we shall consider instead the behaviour of the *mean square temperature*

$$M\left\{\int_0^1 \theta^2\, dx\right\} = \lim_{T \to \infty} \frac{1}{2T} \int_{-T}^{T} \int_0^1 \theta(x, t)^2\, dx\, dt$$

and of the *mean square heat flux*

$$M\left\{\int_0^1 q^2\, dx\right\} = \lim_{T \to \infty} \frac{1}{2T} \int_{-T}^{T} \int_0^1 q(x, t)^2\, dx\, dt,$$

the means being taken with respect to both space and time.

Once again, standard arguments establish that these means exist and equal

$$\tfrac{1}{2}\sum \int_0^1 |F(\omega)\phi(1-x,\omega,a) + G(\omega)\phi(x,\omega,a)|^2\,dx, \tag{3.2.4}$$

$$\tfrac{1}{2}\sum \int_0^1 |-F(\omega)\phi'(1-x,\omega,a) + G(\omega)\phi'(x,\omega,a)|^2\,dx, \tag{3.2.5}$$

respectively.

Our aim is to express the means in terms of means associated with the squares of the sum $f+g$ and the difference $f-g$, and the squares of their derivatives; that is in terms of

$$\begin{aligned} M\{|f \pm g|^2\} &= \tfrac{1}{2}\sum |F(\omega) \pm G(\omega)|^2, \\ M\{|f' \pm g'|^2\} &= \tfrac{1}{2}\sum \omega^2|F(\omega) \pm G(\omega)|^2, \\ M\{|f'' \pm g''|^2\} &= \tfrac{1}{2}\sum \omega^4|F(\omega) \pm G(\omega)|^2, \end{aligned}$$

and so forth.

When the coupling constant $a = 0$, the expressions involve the BERNOULLI *numbers* $B_1, B_2, B_3, \ldots$, which, it will be recalled, can be defined to be the coefficients in the (complex) power series expansion

$$\tfrac{1}{2}z \coth \tfrac{1}{2}z = 1 + \frac{B_1}{2!}z^2 - \frac{B_2}{4!}z^4 + \frac{B_3}{6!}z^6 - \cdots, \qquad |z| < 2\pi. \tag{3.2.6}$$

This definition, which is one among several competitors, ensures that every BERNOULLI number is positive and rational. The first five numbers are

$$B_1 = \tfrac{1}{6}, \quad B_2 = \tfrac{1}{30}, \quad B_3 = \tfrac{1}{42}, \quad B_4 = \tfrac{1}{30}, \quad B_5 = \tfrac{5}{66}.$$

When n is large the numbers become very complicated; their asymptotic behaviour, though, is that indicated by the relation

$$B_n \sim 4\sqrt{\pi n}\left(\frac{n}{e\pi}\right)^{2n} \quad \text{as } n \to \infty.$$

The additional expansion

$$\begin{aligned} \tfrac{1}{2}z \tanh \tfrac{1}{2}z &= \frac{B_1}{2!}(2^2-1)z^2 - \frac{B_2}{4!}(2^4-1)z^4 \\ &\quad + \frac{B_3}{6!}(2^6-1)z^6 - \cdots, \qquad |z| < \pi, \end{aligned} \tag{3.2.7}$$

is a consequence of (3.2.6) and the identity

$$\tanh \tfrac{1}{2}z = 2 \coth z - \coth \tfrac{1}{2}z.$$

We begin by deriving expansions for the mean square temperature and the mean square heat flux which are valid when $a = 0$; this, of course, is the classical case but the result would seem not to have been noticed before its publication in [16].

If we put

$$\begin{aligned} p_0(\omega) &= p(\omega, 0) = (1 + i)\sqrt{\omega/2}, \\ \Theta_0(x, \omega) &= F(\omega)\phi(1 - x, \omega, 0) + G(\omega)\phi(x, \omega, 0) \\ &= (F(\omega) \sinh p_0(1 - x) + G(\omega) \sinh p_0 x) \operatorname{cosech} p_0, \end{aligned}$$

the sum

$$\theta(x, t) = \operatorname{Re} \sum \Theta_0(x, \omega) \exp(i\omega t)$$

is the trigonometric solution of the heat equation

$$\frac{\partial^2 \theta}{\partial x^2} = \frac{\partial \theta}{\partial t}$$

which satisfies the boundary conditions

$$\theta|_{x=0} = f, \qquad \theta|_{x=1} = g.$$

Since

$$\Theta_0'' = i\omega\Theta_0$$

it must be that

$$(\bar{\Theta}_0 \Theta_0')' = i\omega|\Theta_0|^2 + |\Theta_0'|^2$$

and, therefore, it must be that

$$\begin{aligned} i\omega \int_0^1 |\Theta_0|^2\,dx &+ \int_0^1 |\Theta_0'|^2\,dx \\ &= \overline{\Theta_0(1, \omega)}\Theta_0'(1, \omega) - \overline{\Theta_0(0, \omega)}\Theta_0'(0, \omega) \\ &= \overline{G(\omega)}(-F(\omega)p_0 \operatorname{cosech} p_0 + G(\omega)p_0 \coth p_0) \\ &\quad - \overline{F(\omega)}(-F(\omega)p_0 \coth p_0 + G(\omega)p_0 \operatorname{cosech} p_0) \\ &= \tfrac{1}{2}|F(\omega) - G(\omega)|^2 p_0 \coth \tfrac{1}{2}p_0 \\ &\quad + \tfrac{1}{2}|F(\omega) + G(\omega)|^2 p_0 \tanh \tfrac{1}{2}p_0. \end{aligned} \tag{3.2.8}$$

On noting that $p_0^2 = i\omega$ we deduce, from (3.2.6) and (3.2.7), the expansions

$$\tfrac{1}{2}p_0 \coth \tfrac{1}{2}p_0 = 1 + \frac{B_1}{2!}\cdot i\omega + \frac{B_2}{4!}\cdot \omega^2 - \frac{B_3}{6!}\cdot i\omega^3 - \frac{B_4}{8!}\cdot \omega^4 + \cdots,$$

$$\tfrac{1}{2}p_0 \tanh \tfrac{1}{2}p_0 = \frac{B_1}{2!}(2^2 - 1)i\omega + \frac{B_2}{4!}(2^4 - 1)\omega^2$$

$$- \frac{B_3}{6!}(2^6 - 1)i\omega^3 - \frac{B_4}{8!}(2^8 - 1)\omega^4 + \cdots,$$

which converge if $0 < \omega < \pi^2$, and when we substitute these expansions into (3.2.8) and equate real and imaginary parts we obtain the formulae

$$\int_0^1 |\Theta_0|^2\,dx = |F(\omega) - G(\omega)|^2\left(\frac{B_1}{2!} - \frac{B_3}{6!}\,\omega^2 + \frac{B_5}{10!}\,\omega^4 - \cdots\right)$$

$$+ |F(\omega) + G(\omega)|^2\left(\frac{B_1}{2!}(2^2 - 1) - \frac{B_3}{6!}(2^6 - 1)\omega^2\right.$$

$$\left. + \frac{B_5}{10!}(2^{10} - 1)\omega^4 - \cdots\right),$$

$$\int_0^1 |\Theta_0'|^2\,dx = |F(\omega) - G(\omega)|^2\left(1 + \frac{B_2}{4!}\,\omega^2 - \frac{B_4}{8!}\,\omega^4 + \frac{B_6}{12!}\,\omega^6 - \cdots\right)$$

$$+ |F(\omega) + G(\omega)|^2\left(\frac{B_2}{4!}(2^4 - 1)\omega^2 - \frac{B_4}{8!}(2^8 - 1)\omega^4\right.$$

$$\left. + \frac{B_6}{12!}(2^{12} - 1)\omega^6 - \cdots\right).$$

Finally, we sum each side of these formulae over the finite set of exponents ω, and remember that

$$M\left\{\int_0^1 \theta^2\,dx\right\} = \tfrac{1}{2}\sum \int_0^1 |\Theta_0|^2\,dx,$$

$$M\left\{\int_0^1 q^2\,dx\right\} = \tfrac{1}{2}\sum \int_0^1 |\Theta_0'|^2\,dx,$$

and in this way we arrive at

Theorem 10. *If the coupling constant* $a = 0$ *and if each exponent* ω *satisfies* $0 < \omega < \pi^2$, *the mean square temperature is given by the convergent expansion*

$$
\begin{aligned}
M\left\{\int_0^1 \theta^2\,dx\right\} &= \frac{B_1}{2!}M\{(f-g)^2\} - \frac{B_3}{6!}\{(f'-g')^2\} \\
&\quad + \frac{B_5}{10!}M\{(f''-g'')^2\} - \cdots \\
&\quad + \frac{B_1}{2!}(2^2-1)M\{(f+g)^2\} \\
&\quad - \frac{B_3}{6!}(2^6-1)M\{(f'+g')^2\} \\
&\quad + \frac{B_5}{10!}(2^{10}-1)M\{(f''+g'')^2\} - \cdots \\
&= \tfrac{1}{12}M\{(f-g)^2\} - \tfrac{1}{30240}M\{(f'-g')^2\} \\
&\quad + \tfrac{1}{47900160}M\{(f''-g'')^2\} - \cdots \\
&\quad + \tfrac{1}{4}M\{(f+g)^2\} - \tfrac{1}{480}M\{(f'-g')^2\} \\
&\quad + \tfrac{31}{1451520}M\{(f''+g'')^2\} - \cdots,
\end{aligned}
$$

and the mean square heat flux is given by the convergent expansion

$$
\begin{aligned}
M\left\{\int_0^1 q^2\,dx\right\} &= M\{(f-g)^2\} + \frac{B_2}{4!}M\{(f'-g')^2\} \\
&\quad - \frac{B_4}{8!}M\{(f''-g'')^2\} + \cdots \\
&\quad + \frac{B_2}{4!}(2^4-1)M\{(f'+g')^2\} \\
&\quad - \frac{B_4}{8!}(2^8-1)M\{(f''+g'')^2\} + \cdots \\
&= M\{(f-g)^2\} + \tfrac{1}{720}M\{(f'-g')^2\} \\
&\quad - \tfrac{1}{1209600}M\{(f''-g'')^2\} + \cdots \\
&\quad + \tfrac{1}{48}M\{(f'+g')^2\} - \tfrac{17}{80640}M\{(f''+g'')^2\} + \cdots.
\end{aligned}
$$

Closely similar formulae for the mean square temperature and the mean square heat flux continue to hold when the coupling constant is positive, but it is no longer possible to express the coefficients in terms of such a well-studied set of numbers as the BERNOULLI numbers. In what follows we think of the coefficients $F(\omega)$ and $G(\omega)$ which occur in the representations (3.2.2) for the boundary temperatures f and g as being fixed complex numbers and we study what happens when the exponents are small.

Let us write

$$\phi_+ = \phi(1 - x, \omega, a) + \phi(x, \omega, a),$$
$$\phi_- = \phi(1 - x, \omega, a) - \phi(x, \omega, a),$$

so that

$$F(\omega)\phi(1 - x, \omega, a) + G(\omega)\phi(x, \omega, a) = \tfrac{1}{2}(F - G)\phi_- + \tfrac{1}{2}(F + G)\phi_+$$

and

$$\begin{aligned}|F(\omega)\phi(1 - x, \omega, a) &+ G(\omega)\phi(x, \omega, a)|^2 \\ &= \tfrac{1}{4}|F - G|^2|\phi_-|^2 + \tfrac{1}{4}|F + G|^2|\phi_+|^2 \\ &\quad + \tfrac{1}{4}(F - G)\overline{(F + G)}\phi_- \cdot \overline{\phi}_+ + \tfrac{1}{4}\overline{(F - G)}(F + G)\overline{\phi}_- \cdot \phi_+.\end{aligned}$$

Since ϕ_- and ϕ_+ have the properties

$$\phi_-(1 - x, \omega, a) = -\phi_-(x, \omega, a), \qquad \phi_+(1 - x, \omega, a) = \phi_+(x, \omega, a),$$

the integrals

$$\int_0^1 \phi_- \cdot \overline{\phi_+}\, dx, \qquad \int_0^1 \overline{\phi_-} \cdot \phi_+\, dx$$

vanish and, therefore,

$$\begin{aligned}&\int_0^1 |F(\omega)\phi(1 - x, \omega, a) + G(\omega)\phi(x, \omega, a)|^2\, dx \\ &\qquad = \tfrac{1}{4}|F - G|^2 \int_0^1 |\phi_-|^2\, dx + \tfrac{1}{4}|F + G|^2 \int_0^1 |\phi_+|^2\, dx.\end{aligned}$$

A similar argument establishes the equation

$$\begin{aligned}&\int_0^1 |-F(\omega)\phi'(1 - x, \omega, a) + G(\omega)\phi'(x, \omega, a)|^2\, dx \\ &\qquad = \tfrac{1}{4}|F - G|^2 \int_0^1 |\phi'_-|^2\, dx + \tfrac{1}{4}|F + G|^2 \int_0^1 |\phi'_+|^2\, dx.\end{aligned}$$

According to the definition (3.2.3) of ϕ,

$$\phi_- = \frac{\sinh(\frac{1}{2} - x)p}{\sinh \frac{1}{2}p},$$

$$\phi_+ = \frac{p\cosh(\frac{1}{2} - x)p + 2a\sinh\frac{1}{2}p}{p\cosh\frac{1}{2}p + 2a\sinh\frac{1}{2}p},$$

and if we introduce the (real) number $r = \sqrt{(1+a)\omega/2}$ and remember that $p = (1+i)r$ and $p^2 = 2ir^2$ we find, after some calculation, that, as $\omega \to 0$

$$\int_0^1 |\phi_-|^2\,dx = \frac{\sinh r - \sin r}{r(\cosh r - \cos r)}$$

$$= \tfrac{1}{3} - \tfrac{1}{7560}(1+a)^2\omega^2 + o(\omega^2),$$

$$\int_0^1 |\phi_+|^2\,dx$$

$$= \frac{r(\sinh r + \sin r) + 2a(2+a)(\cosh r - \cos r)}{r^2(\cosh r + \cos r) + 2ar(\sinh r + \sin r) + 2a^2(\cosh r - \cos r)}$$

$$= 1 - (\tfrac{1}{120} + \tfrac{1}{360}a)\omega^2 + o(\omega^2),$$

$$\int_0^1 |\phi'_-|^2\,dx = \frac{2r(\sinh r + \sin r)}{\cosh r - \cos r}$$

$$= 4 + \tfrac{1}{180}(1+a)^2\omega^2 + o(\omega^2),$$

$$\int_0^1 |\phi'_+|^2\,dx$$

$$= \frac{2r^3(\sinh r - \sin r)}{r^2(\cosh r + \cos r) + 2ar(\sinh r + \sin r) + 2a^2(\cosh r - \cos r)}$$

$$= \tfrac{1}{12}\omega^2 + o(\omega^2).$$

Thus

$$\int_0^1 |F(\omega)\phi(1-x,\omega,a) + G(\omega)\phi(x,\omega,a)|^2\,dx$$

$$= |F - G|^2(\tfrac{1}{12} - \tfrac{1}{30240}(1+a)^2\omega^2)$$

$$+ |F + G|^2(\tfrac{1}{4} - (\tfrac{1}{480} + \tfrac{1}{1440}a)\omega^2) + o(\omega^2),$$

and

$$\int_0^1 |-F(\omega)\phi'(1-x,\omega,a) + G(\omega)\phi'(x,\omega,a)|^2\,dx$$

$$= |F - G|^2(1 + \tfrac{1}{720}(1+a)^2\omega^2) + \tfrac{1}{48}|F + G|^2\omega^2 + o(\omega^2),$$

and if we sum these asymptotic expressions over the finite set of small positive exponents ω, and recall that the mean square temperature and the mean square heat flux are given by (3.2.4) and (3.2.5) we arrive at

Theorem 11. *If the exponents ω are small, the mean square temperature*

$$M\left\{\int_0^1 \theta^2\,dx\right\}$$

is approximately equal to

$$\tfrac{1}{12}M\{(f-g)^2\} - \tfrac{1}{30240}(1+a)^2 M\{(f'-g')^2\} + \tfrac{1}{4}M\{(f+g)^2\} - (\tfrac{1}{480} + \tfrac{1}{1440}a)M\{(f'+g')^2\},$$

and the mean square heat flux

$$M\left\{\int_0^1 q^2\,dx\right\}$$

is approximately equal to

$$M\{(f-g)^2\} + \tfrac{1}{720}(1+a)^2 M\{(f'-g')^2\} + \tfrac{1}{48}M\{(f'+g')^2\}.$$

When the coupling constant $a = 0$ the terms displayed reduce, as they should, to the corresponding terms in the expansions for the heat equation which were obtained in Theorem 10.

To within an error $o(\omega_{\max}^2)$, where $\omega_{\max}$ is the greatest of the exponents,

$$M\left\{\int_0^1 \theta^2\,dx\right\} - M\left\{\int_0^1 \theta^2\,dx\right\}\bigg|_{a=0} = -\tfrac{1}{30240}(2a+a^2)M\{(f'-g')^2\} - \tfrac{1}{1440}aM\{(f'+g')^2\},$$

and this difference is *negative* unless f and g both vanish identically. To within the same error,

$$M\left\{\int_0^1 q^2\,dx\right\} - M\left\{\int_0^1 q^2\,dx\right\}\bigg|_{a=0} = \tfrac{1}{720}(2a+a^2)M\{(f'-g')^2\},$$

which is *positive* unless f and g coincide. If f and g do coincide we need a more accurate appraisal of the asymptotic behaviour of the integral

$$\int_0^1 |\phi'_+|^2 \, dx.$$

On calculating one additional term in the asymptotic expansion we see that the integral equals

$$\tfrac{1}{12}\omega^2 - (\tfrac{17}{20160} + \tfrac{1}{3360}a + \tfrac{1}{30240}a^2)\omega^4 + o(\omega^4),$$

and we conclude that, when $f = g$,

$$M\left\{\int_0^1 q^2 \, dx\right\} - M\left\{\int_0^1 q^2 \, dx\right\}\Bigg|_{a=0} = -(\tfrac{1}{3360}a + \tfrac{1}{30240}a^2)M\{(f'')^2\},$$

to within an error $o(\omega_{\max}^4)$. This difference is now *negative* and, therefore, we have established the following result which sheds light on the effect of coupling.

Theorem 12. *If the boundary temperatures f and g do not both vanish identically and their exponents are small, the mean square temperature is decreased by coupling, while the mean square heat flux is increased unless $f = g$, in which case it is decreased.*

CHAPTER 4

Approximation by Way of the Heat Equation or the Integro-differential Equation

§4.1. Status of the Heat Equation

At this stage we return to the full thermoelastic equations (1.1.1) and (1.1.2), that is to say to the equations

$$\frac{\partial^2\theta}{\partial x^2}=\frac{\partial\theta}{\partial t}+\sqrt{a}\cdot\frac{\partial^2 u}{\partial x\,\partial t},$$

$$\frac{\partial^2 u}{\partial x^2}=\sqrt{a}\cdot\frac{\partial\theta}{\partial x}+b\frac{\partial^2 u}{\partial t^2},$$

in which the inertial term $b\partial^2u/\partial t^2$ is retained: we suppose, as before, that the faces of the body are clamped and, therefore, that

$$u|_{x=0}=u|_{x=1}=0.$$

These are our original boundary conditions (1.1.3).

In most studies of linear thermoelasticity it is taken for granted that it is permissible to determine the temperature by making the uncoupled approximation, that is by setting the coupling constant $a=0$ in the first of the thermoelastic equations. The implication of this approximation is, of course, that the temperature can be determined by solving the heat equation.

We begin by proving a result [10] which provides some justification for the approximation in the case in which the temperature satisfies the conditions

$$\theta|_{x=0} = 0, \qquad \left.\frac{\partial\theta}{\partial x}\right|_{x=1} = h, \tag{4.1.1}$$

where h is prescribed. These conditions correspond to holding the face $x = 0$ at the reference temperature and subjecting the face $x = 1$ to a known heat flux. It is also possible to justify the approximation [11] when it is the temperature of the face $x = 1$ which is varied in a known way, but the argument is somewhat more troublesome and we shall omit it.

Throughout this section and the next two we take it that:

θ and u are continuous in the closed half-strip $[0, 1] \times [0, \infty)$ and C^4 in the half-strip $[0, 1] \times (0, \infty)$, and they satisfy the thermoelastic equations (1.1.1) *and* (1.1.2), *and the boundary conditions* (1.1.3) *and* (4.1.1).

It will be necessary to restrict the rate at which the heat flux $h(t)$ grows as $t \to \infty$ by requiring that its first and second derivatives h' and h'' be *square-integrable* on $[0, \infty)$; in short the integrals

$$\int_0^\infty (h')^2\, dt, \qquad \int_0^\infty (h'')^2\, dt$$

are required to exist.

The temperature θ is to be compared with an approximating ϕ which satisfies the heat equation

$$\frac{\partial^2\phi}{\partial x^2} = \frac{\partial\phi}{\partial t},$$

and satisfies the same boundary conditions as does θ, namely

$$\phi|_{x=0} = 0, \qquad \left.\frac{\partial\phi}{\partial x}\right|_{x=1} = h.$$

It will be supposed that: *ϕ is continuous in $[0, 1] \times [0, \infty)$ and C^3 in $[0, 1] \times (0, \infty)$.*

Theorem 13. *If h has square-integrable derivatives h' and h'' then $\theta - \phi \to 0$ as $t \to \infty$, the convergence being uniform with respect to x in $[0, 1]$.*

§4.2. Comments on Theorem 13

Some comments on Theorem 13 are in order before we proceed to the proof.

It should be noted, to begin with, that the conclusion holds independently of any prescription of the initial values taken, when $t = 0$, by the temperature θ, the displacement u, the velocity $\partial u/\partial t$, or by the approximating temperature ϕ; it certainly need not be the case that ϕ coincides with θ initially.

The hypotheses on h are satisfied if $h = 0$, that is if the face $x = 0$ is held at the reference temperature and the face $x = 1$ is thermally insulated. Since we can now choose $\phi = 0$ we have deduced a theorem of *asymptotic stability*.

Theorem 14. *If $h = 0$ then, whatever the initial temperature, displacement, or velocity, $\theta \to 0$ as $t \to \infty$, the convergence being uniform with respect to x in* $[0, 1]$.

The same conclusion is true of the displacement,† although in concentrating upon the temperature we have not emphasized the behaviour of the displacement.

The hypotheses upon h are also satisfied if there is an interval $[t_0, \infty)$ on which h is constant. In this case, of course, h must tend to a finite limit as $t \to \infty$, but it is not generally true that the hypotheses force h to tend to a finite limit. Thus, the function $h(t) = \log(1 + t)$ has square-integrable first and second derivatives, and so has the function $h(t) = (1 + t)^\varepsilon$ if $\varepsilon < \frac{1}{2}$.

We might ask if it is possible to lighten the conditions on h, so as to allow it to increase more rapidly for large values of t and yet retain the conclusion that the temperature is approximated by a solution of the heat equation. That no very substantial lightening of the conditions is possible is made clear by the example which follows and which tells us that the exponent in $(1 + t)^\varepsilon$ cannot be increased as far as $\varepsilon = 1$.

Example. *The functions*

$$h(t) = 1 + t,$$

$$\theta(x, t) = \tfrac{1}{6}(1 + a)x^3 - \tfrac{1}{4}ax^2 + \tfrac{1}{2}x + xt,$$

$$u(x, t) = \sqrt{a}\left(\tfrac{1}{24}(1 + a)x^4 - \tfrac{1}{12}ax^3 + \tfrac{1}{4}x^2 - \tfrac{1}{24}(7 - a)x - \tfrac{1}{2}x(1 - x)t\right),$$

$$\phi(x, t) = \tfrac{1}{6}x^3 + \tfrac{1}{2}x + xt,$$

---

† The details of the argument may be found in [10].

satisfy all the hypotheses of Theorem 13, *with the single exception that* h' *is not square-integrable.*

In this example, ϕ does approximate θ in an asymptotic sense, namely

$$\theta(x, t) \sim \phi(x, t) \quad \text{as } t \to \infty,$$

but fails to approximate θ in the more stringent uniform $o(1)$ sense of Theorem 13 because the difference

$$\theta(x, t) - \phi(x, t) = \tfrac{1}{6}ax^3 - \tfrac{1}{4}ax^2$$

is constant in time and does not tend to zero as $t \to \infty$ (except at $x = 0$).

§4.3. Proof of Theorem 13

In order to prove Theorem 13 it will be enough to prove that each of the functions

$$E_1 = \frac{1}{2}\int_0^1 \left(\left(\frac{\partial\theta}{\partial t}\right)^2 + \left(\frac{\partial^2 u}{\partial x\,\partial t}\right)^2 + b\left(\frac{\partial^2 u}{\partial t^2}\right)^2\right) dx,$$

$$F = \frac{1}{2}\int_0^1 \left(\frac{\partial\phi}{\partial t}\right)^2 dx,$$

tends to zero as $t \to \infty$; here E_1 is the analogue of the total energy E which results when θ and u are replaced by their derivatives $\partial\theta/\partial t$ and $\partial u/\partial t$.

To see that these facts about E_1 and F will suffice, let us consider the difference

$$\psi = \theta - \phi,$$

whose second derivative with repect to x is

$$\frac{\partial^2\psi}{\partial x^2} = \frac{\partial^2\theta}{\partial x^2} - \frac{\partial^2\phi}{\partial x^2} = \frac{\partial\theta}{\partial t} + \sqrt{a}\cdot\frac{\partial^2 u}{\partial x\,\partial t} - \frac{\partial\phi}{\partial t}.$$

On integrating with respect to x and taking account of the boundary conditions

$$\psi\,|_{x=0} = \left.\frac{\partial\psi}{\partial x}\right|_{x=1} = 0,$$

we deduce that

$$\psi(x, t) = -\int_0^x y\left(\frac{\partial\theta}{\partial t}(y, t) + \sqrt{a}\cdot\frac{\partial^2 u}{\partial y\,\partial t}(y, t) - \frac{\partial\phi}{\partial t}(y, t)\right) dy$$
$$- x\int_x^1 \left(\frac{\partial\theta}{\partial t}(y, t) + \sqrt{a}\cdot\frac{\partial^2 u}{\partial y\,\partial t}(y, t) - \frac{\partial\phi}{\partial t}(y, t)\right) dy.$$

Thus

$$\max\{|\psi(x, t)| : 0 \leqq x \leqq 1\}$$

does not exceed the integral

$$\int_0^1 \left(\left|\frac{\partial\theta}{\partial t}\right| + \sqrt{a}\cdot\left|\frac{\partial^2 u}{\partial x\,\partial t}\right| + \left|\frac{\partial\phi}{\partial t}\right|\right) dx,$$

which tends to zero if E_1 and F do.

That F tends to zero is a straightforward deduction from the fact that ϕ is a solution of the heat equation

$$\frac{\partial^2\phi}{\partial x^2} = \frac{\partial\phi}{\partial t}$$

and satisfies the boundary conditions

$$\phi|_{x=0} = 0, \qquad \left.\frac{\partial\phi}{\partial x}\right|_{x=1} = h$$

where h' is square-integrable. For, the identity

$$\frac{\partial\phi}{\partial t}\cdot\frac{\partial^2\phi}{\partial t^2} + \left(\frac{\partial^2\phi}{\partial x\,\partial t}\right)^2 = \frac{\partial}{\partial x}\left(\frac{\partial\phi}{\partial t}\cdot\frac{\partial^2\phi}{\partial x\,\partial t}\right)$$

implies that

$$F' + \int_0^1 \left(\frac{\partial^2\phi}{\partial x\,\partial t}\right)^2 dx = h'\cdot\left.\frac{\partial\phi}{\partial t}\right|_{x=1}$$
$$= h'\int_0^1 \frac{\partial^2\phi}{\partial x\,\partial t}\,dx$$
$$\leqq \tfrac{1}{2}h'^2 + \tfrac{1}{2}\int_0^1 \left(\frac{\partial^2\phi}{\partial x\,\partial t}\right)^2 dx$$

and, therefore, that

$$2F' + \int_0^1 \left(\frac{\partial^2\phi}{\partial x\,\partial t}\right)^2 dx \leqq h'^2.$$

Since

$$\left.\frac{\partial\phi}{\partial t}\right|_{x=0} = 0,$$

it must be that

$$\left(\frac{\partial\phi}{\partial t}(x,t)\right)^2 = \left(\int_0^x \frac{\partial^2\phi}{\partial y\,\partial t}(y,t)\,dy\right)^2 \leqq \int_0^1 \left(\frac{\partial^2\phi}{\partial y\,\partial t}(y,t)\right)^2 dy$$

and, on integrating both sides with respect to x over the interval $[0, 1]$, we deduce that

$$2F \leqq \int_0^1 \left(\frac{\partial^2\phi}{\partial x\,\partial t}\right)^2 dx.$$

Thus F satisfies the differential inequality

$$F' + F \leqq \tfrac{1}{2}h'^2,$$

whose right-hand side is integrable on $[0, \infty)$. If we bear in mind that $F \geqq 0$ we can now deduce by means of an elementary argument† that $F(t) \to 0$ as $t \to \infty$, as required.

We turn to the task of showing that $E_1(t) \to 0$ as $t \to \infty$. It is actually a simple matter to prove that E_1 must tend to *some* limit.

† The differential inequality implies that, for each $t \geqq 0$,

$$\begin{aligned} 0 \leqq F(t) &\leqq F(0)\exp(-t) + \tfrac{1}{2}\int_0^t (h'(s))^2 \exp(s-t)\,ds \\ &= F(0)\exp(-t) + \tfrac{1}{2}\int_0^\infty G(s,t)\,ds, \end{aligned}$$

where

$$G(s,t) = \begin{cases} (h'(s))^2\exp(s-t), & 0 \leqq s \leqq t, \\ 0, & t < s < \infty. \end{cases}$$

For each fixed s in $[0, \infty)$, $G(s, t) \to 0$ as $t \to \infty$ and, for all values of s and t in $[0, \infty)$, $0 \leqq G(s, t) \leqq (h'(s))^2$. LEBESGUE's Theorem of Dominated Convergence now enables us to conclude that

$$\int_0^\infty G(s,t)\,ds \to 0 \quad \text{as } t \to \infty$$

and, hence, that $F(t) \to 0$.

For, Theorem 1 tells us that, in the present context, the total energy E satisfies

$$E' + \int_0^1 \left(\frac{\partial\theta}{\partial x}\right)^2 dx = h\theta|_{x=1}$$

$$= h\int_0^1 \frac{\partial\theta}{\partial x}\, dx$$

$$\leqq \tfrac{1}{2}h^2 + \tfrac{1}{2}\int_0^1 \left(\frac{\partial\theta}{\partial x}\right)^2 dx,$$

and, therefore,

$$2E' \leqq h^2.$$

On repeating this argument, but with θ, u, and h, replaced by their derivatives with respect to t, we arrive at the inequality

$$2E_1' \leqq h'^2$$

for the derivative of E_1. Since

$$\frac{d}{dt}\left(2E_1(t) + \int_t^\infty (h'(s))^2\, ds\right) \equiv 2E_1'(t) - (h'(t))^2 \leqq 0$$

the nonnegative function

$$2E_1(t) + \int_t^\infty (h'(s))^2\, ds$$

must decrease and, hence, must tend to a limit as $t \to \infty$. Thus E_1 itself must tend to a limit.

To show that E_1 tends to zero we establish—what is the most difficult part of the argument—that E_1 is integrable on $[0, \infty)$. We begin by deriving appropriate estimates on the integral

$$\int_0^T E\, dt, \tag{4.3.1}$$

where T is any large number, and then we make the switch from E to E_1 by considering derivatives.

The integral (4.3.1) can be estimated in terms of $E(T)$ and certain integrals which involve θ only, as is done in the inequality (4.3.2) which we proceed to derive. For, the thermoelastic equation (1.1.2) implies the identity

$$\left(\frac{\partial u}{\partial x}\right)^2 + \left(\frac{\partial u}{\partial x} - \sqrt{a}\cdot\theta\right)^2 = 2\frac{\partial}{\partial x}\left(u\left(\frac{\partial u}{\partial x} - \sqrt{a}\cdot\theta\right)\right)$$
$$-2b\frac{\partial}{\partial t}\left(u\frac{\partial u}{\partial t}\right) + 2b\left(\frac{\partial u}{\partial t}\right)^2 + a\theta^2,$$

and if we integrate with respect to x and remember that $u|_{x=0} = u|_{x=1} = 0$ we deduce that

$$\int_0^1 \left(\frac{\partial u}{\partial x}\right)^2 dx \leqq -2b\frac{d}{dt}\int_0^1 u\frac{\partial u}{\partial t}\,dx + 2b\int_0^1 \left(\frac{\partial u}{\partial t}\right)^2 dx + a\int_0^1 \theta^2\,dx.$$

On adding the terms

$$\int_0^1 \theta^2\,dx + b\int_0^1 \left(\frac{\partial u}{\partial t}\right)^2 dx$$

to each side we conclude that

$$2E \leqq -2b\frac{d}{dt}\int_0^1 u\frac{\partial u}{\partial t}\,dx + 3b\int_0^1 \left(\frac{\partial u}{\partial t}\right)^2 dx + (1+a)\int_0^1 \theta^2\,dx$$

and, therefore,

$$2\int_0^T E\,dt \leqq 2b\int_0^1 u\frac{\partial u}{\partial t}\,dx\bigg|_{t=0} - 2b\int_0^1 u\frac{\partial u}{\partial t}\,dx\bigg|_{t=T}$$
$$+ 3b\int_0^T\int_0^1 \left(\frac{\partial u}{\partial t}\right)^2 dx\,dt + (1+a)\int_0^T\int_0^1 \theta^2\,dx\,dt.$$

The fact that $u|_{x=0} = 0$ enables us to replace

$$\int_0^1 u^2\,dx$$

by the larger integral

$$\int_0^1 \left(\frac{\partial u}{\partial x}\right)^2 dx$$

and, hence, to deduce that

$$\begin{aligned}2b\left|\int_0^1 u\frac{\partial u}{\partial t}\,dx\right| &\leqq b\int_0^1\left(u^2+\left(\frac{\partial u}{\partial t}\right)^2\right)dx\\ &\leqq b\int_0^1\left(\left(\frac{\partial u}{\partial x}\right)^2+\left(\frac{\partial u}{\partial t}\right)^2\right)dx\\ &\leqq 2(1+b)E,\end{aligned}$$

and that

$$\begin{aligned}2\int_0^T E\,dt \leqq 2(1+b)(E(0)+E(T)) &+ 3b\int_0^T\int_0^1\left(\frac{\partial u}{\partial t}\right)^2 dx\,dt\\ &+(1+a)\int_0^T\int_0^1 \theta^2\,dx\,dt.\end{aligned}$$

On the other hand, if we integrate the thermoelastic equation (1.1.1) with respect to x and appeal to the boundary conditions on the face $x = 1$ we deduce that

$$\sqrt{a}\cdot\frac{\partial u}{\partial t}(x,t) = -h(t)+\frac{\partial\theta}{\partial x}(x,t)+\int_x^1\frac{\partial\theta}{\partial t}(y,t)\,dy,$$

that

$$\tfrac{1}{3}a\left(\frac{\partial u}{\partial t}(x,t)\right)^2 \leqq h(t)^2+\left(\frac{\partial\theta}{\partial x}(x,t)\right)^2+\int_0^1\left(\frac{\partial\theta}{\partial t}(y,t)\right)^2 dy,$$

and, therefore,

$$\tfrac{1}{3}a\int_0^1\left(\frac{\partial u}{\partial t}\right)^2 dx \leqq h^2+\int_0^1\left(\left(\frac{\partial\theta}{\partial x}\right)^2+\left(\frac{\partial\theta}{\partial t}\right)\right)dx.$$

Accordingly, we have proved the inequality

$$\begin{aligned}2\int_0^T E\,dt &\leqq 2(1+b)(E(0)+E(T))\\ &\quad+\frac{9b}{a}\left(\int_0^T h^2\,dt+\int_0^T\int_0^1\left(\left(\frac{\partial\theta}{\partial x}\right)^2+\left(\frac{\partial\theta}{\partial t}\right)^2\right)dx\,dt\right)\\ &\quad+(1+a)\int_0^T\int_0^1\theta^2\,dx\,dt,\end{aligned}$$

and, hence, the required inequality

$$2\int_0^T E\,dt \leqq 2(1+b)(E(0)+E(T)) + \frac{9b}{a}\int_0^T h^2\,dt$$
$$+\left(1+a+\frac{9b}{a}\right)\int_0^T\int_0^1\left(\frac{\partial\theta}{\partial x}\right)^2 dx\,dt$$
$$+\frac{9b}{a}\int_0^T\int_0^1\left(\frac{\partial^2\theta}{\partial x\,\partial t}\right)^2 dx\,dt \qquad (4.3.2)$$

follows by a further appeal to the boundary condition $\theta|_{x=0}=0$.

Finally, we recall that E satisfies the inequality

$$2E' + \int_0^1\left(\frac{\partial\theta}{\partial x}\right)^2 dx \leqq h^2$$

and that E_1 satisfies the inequality

$$2E_1' + \int_0^1\left(\frac{\partial^2\theta}{\partial x\,\partial t}\right)^2 dx\,dt \leqq h'^2.$$

Integration with respect to t yields the estimates

$$2E(T) + \int_0^T\int_0^1\left(\frac{\partial\theta}{\partial x}\right)^2 dx\,dt \leqq 2E(0) + \int_0^T h^2\,dt, \qquad (4.3.3)$$

$$\int_0^T\int_0^1\left(\frac{\partial^2\theta}{\partial x\,\partial t}\right)^2 dx\,dt \leqq 2E_1(0) + \int_0^T h'^2\,dt, \qquad (4.3.4)$$

where the nonnegative term $2E_1(T)$ has been omitted from the left-hand side of (4.3.4).

On using (4.3.3) and (4.3.4) to estimate the right-hand side of (4.3.2) we conclude that

$$\int_0^T E\,dt$$

is bounded above by a constant multiple of

$$1 + \int_0^T (h^2 + h'^2)\,dt.$$

In the same way, the integral

$$\int_0^T E_1\,dt$$

is bounded above by a constant multiple of

$$1 + \int_0^T (h'^2 + h''^2)\, dt$$

and, since h' and h'' are square-integrable on $[0, \infty)$, we conclude that E_1 is integrable on $[0, \infty)$. This completes the proof of Theorem 13.

§4.4. Mean and Recurrence Properties of the Temperature

The methods of Section 4.3 can be used to establish further results concerning the relationship between FOURIER's theory and the dynamic thermoelastic theory† in which the effects of both coupling and inertia are retained.

We shall suppose, as before, that θ and u satisfy the thermoelastic equations and the boundary conditions

$$\theta|_{x=0} = 0, \qquad \left.\frac{\partial \theta}{\partial x}\right|_{x=1} = h,$$

$$u|_{x=0} = u|_{x=1} = 0,$$

but on this occasion h will be subject to a different type of restriction.

We intend to show that, if $h(t)$ does not grow too rapidly as $t \to \infty$, the existence of the *mean heat flux*‡

$$M\{h\} = \lim_{T\to\infty} \frac{1}{T} \int_0^T h(t)\, dt$$

implies the existence of the *pointwise mean temperature*

$$M\{\theta, x\} = \lim_{T\to\infty} \frac{1}{T} \int_0^T \theta(x, t)\, dt$$

throughout the unit interval; *we allow the possibility that these means may take the values* $\pm\infty$.

---

† See [14], where corresponding results are proved for three-dimensional bodies.

‡ Since θ and u are defined only on the half-strip $[0, 1] \times [0, \infty)$ the definition of the mean value operator M differs slightly from that of Chapter 3.

When $M\{h\}$ is finite it turns out that the pointwise mean temperature is

$$M\{\theta, x\} = xM\{h\},$$

and this, of course, is just the solution of the equilibrium boundary value problem

$$\left.\begin{aligned} &\frac{d^2}{dx^2} M\{\theta, x\} = 0, \\ M\{\theta, 0\} = 0, \qquad &\frac{d}{dx} M\{\theta, 1\} = M\{h\}. \end{aligned}\right\} \tag{4.4.1}$$

Theorem 15. *Suppose that h satisfies the growth condition*

$$\int_0^T h(t)^2 \, dt = o(T^2) \quad \text{as } T \to \infty,$$

and that the mean value $M\{h\}$ exists.

Then, if $M\{h\}$ is finite, $M\{\theta, x\}$ exists, and is finite, at each point x in $[0, 1]$ and

$$M\{\theta, x\} = xM\{h\}.$$

If $M\{h\}$ is not finite, $M\{\theta, 0\} = 0$ but if x is any point of $(0, 1]$, $M\{\theta, x\} = +\infty$ or $-\infty$ according as $M\{h\} = +\infty$ or $-\infty$.

The growth condition is satisfied, and the mean $M\{h\}$ exists and is finite, if, for example, h is a periodic function or a uniformly almost periodic function. The function $h(t) = (1 + t)^{1/3}$ is one for which the growth condition is satisfied and $M\{h\}$ exists and is equal to $+\infty$.

In order to prove Theorem 15 we recall that, as we established in the course of proving Theorem 13, the derivative of the total energy satisfies

$$2E' \leqq h^2$$

and, therefore,

$$2E(T) \leqq 2E(0) + \int_0^T h(t)^2 \, dt$$

for every positive T. Thus, the growth condition on h ensures that

$$E(T) = o(T^2).$$

We recall too that we can write the thermoelastic equation (1.1.1) as

$$\frac{\partial^2 \theta}{\partial x^2} = \frac{\partial \eta}{\partial t},$$

where

$$\eta = \theta + \sqrt{a} \cdot \frac{\partial u}{\partial x}$$

is the entropy.

Now let us put

$$\xi(x, T) = \frac{1}{T} \int_0^T (\theta(x, t) - xh(t))\, dt.$$

The second derivative of ξ with respect to x is

$$\frac{\partial^2 \xi}{\partial x^2}(x, T) = \frac{1}{T} \int_0^T \frac{\partial^2 \theta}{\partial x^2}(x, t)\, dt = \frac{1}{T}(\eta(x, T) - \eta(x, 0)),$$

and ξ satisfies the boundary conditions

$$\xi|_{x=0} = \left.\frac{\partial \xi}{\partial x}\right|_{x=1} = 0.$$

Thus

$$\xi(x, T) = \frac{1}{T} \int_0^x y(\eta(y, 0) - \eta(y, T))\, dy + \frac{1}{T} \cdot x \int_x^1 (\eta(y, 0) - \eta(y, T))\, dy$$

and

$$\begin{aligned} |\xi(x, T)| &\leqq \frac{1}{T} \int_0^1 (|\eta(y, 0)| + |\eta(y, T)|)\, dy \\ &\leqq \frac{1}{T} \sqrt{\int_0^1 |\eta(y, 0)|^2\, dy} + \frac{1}{T} \sqrt{\int_0^1 |\eta(y, T)|^2\, dy}. \end{aligned}$$

On appealing to the inequalities

$$\begin{aligned} \eta^2 &\leqq (1 + a)\left(\theta^2 + \left(\frac{\partial u}{\partial x}\right)^2\right) \\ &\leqq (1 + a)\left(\theta^2 + \left(\frac{\partial u}{\partial x}\right)^2 + b\left(\frac{\partial u}{\partial t}\right)^2\right) = 2(1 + a)e, \end{aligned}$$

where e is the total energy density, we conclude that

$$|\xi(x, T)| \leqq \frac{1}{T}\sqrt{2(1+a)}(\sqrt{E(0)} + \sqrt{E(T)}) = o(1).$$

In short, we have proved the order relation

$$\frac{1}{T}\int_0^T \theta(x, t)\, dt = x \cdot \frac{1}{T}\int_0^T h(t)\, dt + o(1),$$

which is sufficient to establish the theorem.

There are two immediate corollaries. The first describes a recurrence property of the temperature and says that, when the mean $M\{h\}$ is finite, the temperature comes arbitrarily close to the solution of the equilibrium boundary value problem (4.4.1) at arbitrarily large times. The second says that, when the mean $M\{h\} = \pm\infty$, the temperature is arbitrarily large in magnitude at arbitrarily large times (except at $x = 0$).

Theorem 16. *Suppose that h satisfies the hypotheses of Theorem* 15 *and that $M\{h\}$ is finite. Let x be any point of* $[0, 1]$ *and let ε and T be positive numbers which are arbitrarily small and arbitrarily large, respectively. Then there is a number t, which depends upon x, ε, and T, such that $t > T$ and*

$$|\theta(x, t) - xM\{h\}| < \varepsilon.$$

Theorem 17. *Suppose that h satisfies the hypotheses of Theorem* 15 *and that $M\{h\} = \pm\infty$. Let x be any point of* $(0, 1]$, *the point $x = 0$ being excluded, and let N and T be arbitrarily large positive numbers. Then there is a number t, which depends upon x, N, and T, such that $t > T$ and*

$$\theta(x, t) > N \quad \text{or} \quad \theta(x, t) < -N,$$

according as $M\{h\}$ equals $+\infty$ or $-\infty$.

§4.5. Status of the Integro-differential Equation

We retain the hypothesis that: *θ and u are continuous in* $[0, 1] \times [0, \infty)$, *and C^4 in* $[0, 1] \times (0, \infty)$, *and they satisfy the thermoelastic equations* (1.1.1) *and* (1.1.2), *and the boundary conditions* (1.1.3) *and* (4.1.1).

In this section, however, θ will be compared with a ϕ which is a solution, not of the heat equation, but of the integro-differential equation

$$\frac{\partial^2 \phi}{\partial x^2}(x, t) = (1 + a)\frac{\partial \phi}{\partial t}(x, t) - a\frac{d}{dt}\int_0^1 \phi(y, t)\, dy,$$

and satisfies the same boundary conditions as does θ, namely

$$\phi|_{x=0} = 0, \qquad \left.\frac{\partial \phi}{\partial x}\right|_{x=1} = h.$$

We shall require that: *ϕ is continuous in* $[0, 1] \times [0, \infty)$ *and* C^6 *in* $[0, 1] \times (0, \infty)$.

The derivation of the integro-differential equation takes account of the effect of coupling but ignores the effect of inertia, whereas the standard derivation of the heat equation ignores both effects. It might be surmised, therefore, that the integro-differential equation would provide approximations to the temperature under conditions on the growth of h which are less stringent than those required by Theorem 13. That this is so is shown by our next theorem.†

Theorem 18. *If h has square-integrable derivatives h'', h''', h'''' (of orders* 2, 3, 4), *then $\theta - \phi \to 0$ as $t \to \infty$, the convergence being uniform with respect to x in* $[0, 1]$.

Once again, the result is true independently of the initial values taken by θ, u, $\partial u/\partial t$, or the approximating ϕ.

The hypotheses on h are satisfied if, for example, $h(t) = (1 + t)^\varepsilon$ where $\varepsilon < \frac{3}{2}$ and in particular, *linear* growth of h, that is $h(t) = 1 + t$, lies within the scope of approximation by way of the integro-differential equation; it will be recalled, of course, that the example constructed in Section 4.2 shows that the heat equation fails to provide a uniform $o(1)$ approximation when $h(t) = 1 + t$.

Approximation by way of the integro-differential equation must itself fail when the exponent in $h(t) = (1 + t)^\varepsilon$ is increased to $\varepsilon = 3$; in that case we should have to take account of the inertial term $b\, \partial^2 u/\partial t^2$ if we desired to approximate θ in the uniform $o(1)$ sense. We can see that this is so by considering an example which is similar to, but more elaborate than, that constructed earlier.

† See [17].

Example. *There are unique polynomials* $p_0(x, a, b)$, $p_1(x, a)$, $p_2(x, a)$, $p_3(x, a, b)$, $p_4(x, a, b)$, $p_5(x, a)$ *such that the functions*

$$h(t) = (1 + t)^3,$$
$$\theta(x, t) = p_0(x, a, b) + p_1(x, a)t + p_2(x, a)t^2 + xt^3,$$
$$u(x, t) = \sqrt{a}(p_3(x, a, b) + p_4(x, a, b)t + p_5(x, a)t^2 - \tfrac{1}{2}x(1 - x)t^3),$$
$$\phi(x, t) = p_0(x, a, 0) + p_1(x, a)t + p_2(x, a)t^2 + xt^3,$$

(this last being obtained by setting $b = 0$ *in the expression for* θ*) satisfy all the hypotheses of Theorem* 18, *except that* h'' *and* h''' *are not square-integrable.*

The determination of the polynomials involved is a straightforward, if tedious, matter. The only one of which we require detailed knowledge is

$$\begin{aligned} p_0(x, a, b) = {} & \tfrac{1}{840}(1 + a)^3x^7 - \tfrac{1}{240}a(1 + a)^2x^6 + \tfrac{1}{40}(1 + a)^2x^5 \\ & - \tfrac{1}{96}a(1 + a)(7 - a)x^4 + \tfrac{1}{24}(1 + a)(5 + a)x^3 \\ & - \tfrac{1}{240}a(91 - 6a + 3a^2)x^2 + \tfrac{1}{120}(29 + 6a - 3a^2)x \\ & + \tfrac{1}{40}ab(2x^5 - 5x^4 + 5x^2). \end{aligned}$$

As in our previous example, it happens that

$$\theta(x, t) \sim \phi(x, t) \quad \text{as } t \to \infty,$$

but the difference

$$\theta(x, t) - \phi(x, t) = p_0(x, a, b) - p_0(x, a, 0) = \tfrac{1}{40}ab(2x^5 - 5x^4 + 5x^2)$$

does not tend to zero as $t \to \infty$ (except at $x = 0$) and, therefore, we cannot relax the hypotheses on h so as to include the case $h(t) = (1 + t)^3$.

In order to prove Theorem 18 we begin by considering the behaviour of

$$E_0 = \tfrac{1}{2}(1 + a)\int_0^1 \phi^2\,dx - \tfrac{1}{2}a\left(\int_0^1 \phi\,dx\right)^2,$$

which, as a comparison with (2.1.4) shows, is the total energy associated with the approximating ϕ. Theorem 1 tells us that the derivative of E_0 is given by the equation

$$E_0' + \int_0^1 \left(\frac{\partial \phi}{\partial x}\right)^2 dx = h\phi|_{x=1}$$

and, since

$$h\phi|_{x=1} = h\int_0^1 \frac{\partial\phi}{\partial x}\,dx \leqq \tfrac{1}{2}h^2 + \tfrac{1}{2}\int_0^1 \left(\frac{\partial\phi}{\partial x}\right)^2 dx,$$

we see that

$$2E_0' + \int_0^1 \left(\frac{\partial\phi}{\partial x}\right)^2 dx \leqq h^2.$$

On integrating with respect to t over an interval $[0, T]$ and omitting a nonnegative term from the left-hand side, we see that

$$\int_0^T \int_0^1 \left(\frac{\partial\phi}{\partial x}\right)^2 dx\,dt \leqq 2E_0(0) + \int_0^T h^2\,dt$$

for every positive T. We obtain similar relations if we replace ϕ and h by their second, third, and fourth, derivatives with respect to t and, since we have supposed h'', h''', h'''' to be square-integrable, we have deduced that the integrals

$$\int_0^\infty \int_0^1 \left(\frac{\partial^{n+1}\phi}{\partial x\,\partial t^n}\right)^2 dx\,dt \qquad (n = 2, 3, 4) \tag{4.5.1}$$

must converge.

Next we introduce the function

$$v(x, t) = \sqrt{a}\cdot(1 - x)\int_0^x \phi(y, t)\,dy - \sqrt{a}\cdot x\int_x^1 \phi(y, t)\,dy$$

which satisfies

$$\int_0^1 v^2\,dx \leqq a\int_0^1 \phi^2\,dx$$

and, in view of the boundary condition $\phi|_{x=0} = 0$, must also satisfy the inequality

$$\int_0^1 v^2\,dx \leqq a\int_0^1 \left(\frac{\partial\phi}{\partial x}\right)^2 dx.$$

Similar inequalities hold, with v and ϕ replaced by their second, third, and fourth, derivatives with respect to t and, thus, the convergence of the integrals (4.5.1) ensures the convergence of the integrals

$$\int_0^\infty \int_0^1 \left(\frac{\partial^n v}{\partial t^n}\right)^2 dx\,dt \qquad (n = 2, 3, 4). \tag{4.5.2}$$

At this stage we note that the definition of v ensures that

$$v|_{x=0} = v|_{x=1} = 0,$$

and that

$$\frac{\partial v}{\partial x}(x, t) = \sqrt{a}\left(\phi(x, t) - \int_0^1 \phi(y, t)\, dy\right),$$

$$\frac{\partial^2 v}{\partial x^2} = \sqrt{a} \cdot \frac{\partial \phi}{\partial x}.$$

Thus, we can rewrite the integro-differential equation that ϕ satisfies as the equation

$$\frac{\partial^2 \phi}{\partial x^2} = \frac{\partial \phi}{\partial t} + \sqrt{a} \cdot \frac{\partial^2 v}{\partial x\, \partial t}$$

and, bearing in mind that θ and u satisfy the thermoelastic equations (1.1.1) and (1.1.2), we conclude that the differences

$$\psi = \theta - \phi, \qquad w = u - v,$$

satisfy the equations

$$\frac{\partial^2 \psi}{\partial x^2} = \frac{\partial \psi}{\partial t} + \sqrt{a} \cdot \frac{\partial^2 w}{\partial x\, \partial t}, \tag{4.5.3}$$

$$\frac{\partial^2 w}{\partial x^2} = \sqrt{a} \cdot \frac{\partial \psi}{\partial x} + b\frac{\partial^2 w}{\partial t^2} + b\frac{\partial^2 v}{\partial t^2}, \tag{4.5.4}$$

and the homogeneous boundary conditions

$$\psi|_{x=0} = \left.\frac{\partial \psi}{\partial x}\right|_{x=1} = w|_{x=0} = w|_{x=1} = 0. \tag{4.5.5}$$

On integrating equation (4.5.3) with respect to x and taking account of the conditions on the face $x = 1$ we find that

$$\sqrt{a} \cdot \frac{\partial w}{\partial t}(x, t) = \frac{\partial \psi}{\partial x}(x, t) + \frac{d}{dt}\int_x^1 \psi(y, t)\, dy. \tag{4.5.6}$$

This last equation will enable us to deduce a formula (4.5.7) which is the key to the proof; it asserts that the derivative of the nonnegative function

$$F(t) = \frac{1}{2}\int_0^1 \left(\left(\psi(x,t) + \frac{b}{\sqrt{a}}\int_0^x \frac{\partial^2 v}{\partial t^2}(y,t)\,dy\right)^2 + \left(\frac{\partial w}{\partial x}(x,t)\right)^2 + b\left(\frac{\partial w}{\partial t}(x,t)\right)^2\right) dx$$

is given by

$$\begin{aligned} F' + \int_0^1 \left(\frac{\partial \psi}{\partial x}\right)^2 dx &= -\frac{b}{\sqrt{a}}\int_0^1 \frac{\partial \psi}{\partial x}\cdot\frac{\partial^2 v}{\partial t^2}\,dx + \frac{b}{\sqrt{a}}\int_0^1 \frac{\partial^3 v}{\partial t^3}(x,t)\,dx \int_x^1 \psi(y,t)\,dy \\ &\quad + \frac{b^2}{a}\int_0^1 dx \int_0^x \frac{\partial^2 v}{\partial t^2}(y,t)\,dy \int_0^x \frac{\partial^3 v}{\partial t^3}(z,t)\,dz. \end{aligned} \tag{4.5.7}$$

From our point of view, (4.5.7) has the advantage that w is absent from the right-hand side.

To verify (4.5.7) we note that

$$\begin{aligned} F' + \int_0^1 \left(\frac{\partial \psi}{\partial x}\right)^2 dx &= \int_0^1 \left[\left(\psi(x,t) + \frac{b}{\sqrt{a}}\int_0^x \frac{\partial^2 v}{\partial t^2}(y,t)\,dy\right)\right. \\ &\quad \times \left(\frac{\partial \psi}{\partial t}(x,t) + \frac{b}{\sqrt{a}}\int_0^x \frac{\partial^3 v}{\partial t^3}(y,t)\,dy\right) \\ &\quad \left. + \frac{\partial w}{\partial x}\cdot\frac{\partial^2 w}{\partial x\,\partial t} + b\frac{\partial w}{\partial t}\cdot\frac{\partial^2 w}{\partial t^2}\right] dx + \int_0^1 \left(\frac{\partial \psi}{\partial x}\right)^2 dx, \end{aligned}$$

and we rearrange the right-hand side as the sum

$$\begin{aligned} &\int \left(\psi\cdot\frac{\partial \psi}{\partial t} + \left(\frac{\partial \psi}{\partial x}\right)^2 + \frac{\partial w}{\partial x}\cdot\frac{\partial^2 w}{\partial x\,\partial t} + b\frac{\partial w}{\partial t}\cdot\frac{\partial^2 w}{\partial t^2}\right) dx \\ &\quad + \frac{b}{\sqrt{a}}\cdot\frac{d}{dt}\int_0^1 \psi(x,t)\,dx \int_0^x \frac{\partial^2 v}{\partial t^2}(y,t)\,dy \\ &\quad + \frac{b^2}{a}\int_0^1 dx \int_0^x \frac{\partial^2 v}{\partial t^2}(y,t)\,dy \int_0^x \frac{\partial^3 v}{\partial t^3}(z,t)\,dz. \end{aligned}$$

Because of (4.5.3), (4.5.4), and (4.5.5), the first of the integrals in this sum equals

$$\int_0^1 \left[\psi\left(\frac{\partial^2 \psi}{\partial x^2} - \sqrt{a}\cdot\frac{\partial^2 w}{\partial x\,\partial t}\right) + \left(\frac{\partial \psi}{\partial x}\right)^2 + \frac{\partial w}{\partial x}\cdot\frac{\partial^2 w}{\partial x\,\partial t} \right.$$

$$\left. + \frac{\partial w}{\partial t}\left(\frac{\partial^2 w}{\partial x^2} - \sqrt{a}\cdot\frac{\partial \psi}{\partial x} - b\frac{\partial^2 v}{\partial t^2}\right)\right] dx$$

$$= \int_0^1 \left[\frac{\partial}{\partial x}\left(\psi\frac{\partial \psi}{\partial x} - \sqrt{a}\cdot\psi\frac{\partial w}{\partial t} + \frac{\partial w}{\partial x}\cdot\frac{\partial w}{\partial t}\right) - b\frac{\partial w}{\partial t}\cdot\frac{\partial^2 v}{\partial t^2}\right] dx$$

$$= -b\int_0^1 \frac{\partial w}{\partial t}\cdot\frac{\partial^2 v}{\partial t^2}\,dx,$$

and, in view of (4.5.6), this in turn equals

$$-\frac{b}{\sqrt{a}}\cdot\int_0^1 \frac{\partial \psi}{\partial x}\cdot\frac{\partial^2 v}{\partial t^2}\,dx - \frac{b}{\sqrt{a}}\int_0^1 \frac{\partial^2 v}{\partial t^2}(x, t)\,dx \int_x^1 \frac{\partial \psi}{\partial t}(y, t)\,dy$$

$$= -\frac{b}{\sqrt{a}}\cdot\int_0^1 \frac{\partial \psi}{\partial x}\cdot\frac{\partial^2 v}{\partial t^2}\,dx - \frac{b}{\sqrt{a}}\cdot\frac{d}{dt}\int_0^1 \frac{\partial^2 v}{\partial t^2}(x, t)\,dx \int_x^1 \psi(y, t)\,dy$$

$$+ \frac{b}{\sqrt{a}}\int_0^1 \frac{\partial^3 v}{\partial t^3}(x, t)\,dx \int_x^1 \psi(y, t)\,dy$$

$$= -\frac{b}{\sqrt{a}}\cdot\int_0^1 \frac{\partial \psi}{\partial x}\cdot\frac{\partial^2 v}{\partial t^2}\,dx - \frac{b}{\sqrt{a}}\cdot\frac{d}{dt}\int_0^1 \psi(x, t)\,dx \int_0^x \frac{\partial^2 v}{\partial t^2}(y, t)\,dy$$

$$+ \frac{b}{\sqrt{a}}\int_0^1 \frac{\partial^3 v}{\partial t^3}(x, t)\,dx \int_x^1 \psi(y, t)\,dy$$

and, hence, (4.5.7) is correct.

The boundary condition $\psi|_{x=0} = 0$ ensures that

$$\int_0^1 \psi^2\,dx \leqq \int_0^1 \left(\frac{\partial \psi}{\partial x}\right)^2 dx$$

and, in consequence, the right-hand side of (4.5.7) does not exceed

$$
\begin{aligned}
&\frac{b}{\sqrt{a}} \cdot \sqrt{\int_0^1 \left(\frac{\partial\psi}{\partial x}\right)^2 dx \cdot \int_0^1 \left(\frac{\partial^2 v}{\partial t^2}\right)^2 dx} \\
&\qquad + \frac{b}{\sqrt{a}} \cdot \sqrt{\int_0^1 \left(\frac{\partial\psi}{\partial x}\right)^2 dx \cdot \int_0^1 \left(\frac{\partial^3 v}{\partial t^3}\right)^2 dx} \\
&\qquad + \frac{b^2}{a} \sqrt{\int_0^1 \left(\frac{\partial^2 v}{\partial t^2}\right)^2 dx \cdot \int_0^1 \left(\frac{\partial^3 v}{\partial t^3}\right)^2 dx} \\
&\qquad \leqq \tfrac{1}{4} \int_0^1 \left(\frac{\partial\psi}{\partial x}\right)^2 dx + \frac{b^2}{a} \int_0^1 \left(\frac{\partial^2 v}{\partial t^2}\right)^2 dx \\
&\qquad + \tfrac{1}{4} \int_0^1 \left(\frac{\partial\psi}{\partial x}\right)^2 dx + \frac{b^2}{a} \int_0^1 \left(\frac{\partial^3 v}{\partial t^3}\right)^2 dx \\
&\qquad + \frac{b^2}{2a} \int_0^1 \left(\frac{\partial^2 v}{\partial t^2}\right)^2 dx + \frac{b^2}{2a} \int_0^1 \left(\frac{\partial^3 v}{\partial t^3}\right)^2 dx \\
&\qquad = \tfrac{1}{2} \int_0^1 \left(\frac{\partial\psi}{\partial x}\right)^2 dx + \frac{3b^2}{2a} \int_0^1 \left[\left(\frac{\partial^2 v}{\partial t^2}\right)^2 + \left(\frac{\partial^3 v}{\partial t^3}\right)^2\right] dx.
\end{aligned}
$$

Thus, (4.5.7) implies that

$$
2F' + \int_0^1 \left(\frac{\partial\psi}{\partial x}\right)^2 dx \leqq \frac{3b^2}{a} \int_0^1 \left[\left(\frac{\partial^2 v}{\partial t^2}\right)^2 + \left(\frac{\partial^3 v}{\partial t^3}\right)^2\right] dx
$$

and, upon integrating with respect to t, we deduce that

$$
\int_0^T \int_0^1 \left(\frac{\partial\psi}{\partial x}\right)^2 dx\, dt \leqq 2F(0) + \frac{3b^2}{a} \int_0^T \int_0^1 \left[\left(\frac{\partial^2 v}{\partial t^2}\right)^2 + \left(\frac{\partial^3 v}{\partial t^3}\right)^2\right] dx\, dt
$$

for every positive T. Since the integrals (4.5.2) converge when $n = 2$ and $n = 3$ it must be the case that the integral

$$
\int_0^\infty \int_0^1 \left(\frac{\partial\psi}{\partial x}\right)^2 dx\, dt
$$

converges.

If now we repeat our argument, with θ, u, ϕ, v, ψ, w, and h, replaced by their derivatives with respect to t, we can deduce in just the same way that the integral

$$
\int_0^\infty \int_0^1 \left(\frac{\partial^2\psi}{\partial x\, \partial t}\right)^2 dx\, dt
$$

converges, and it follows, by a standard argument, that

$$\int_0^1 \left(\frac{\partial \psi}{\partial x}\right)^2 dx \to 0 \quad \text{as } t \to \infty.$$

Since $\psi|_{x=0} = 0$ we conclude, as required, that $\psi \to 0$ as $t \to \infty$, the convergence being uniform with respect to x on the interval $[0, 1]$.

CHAPTER 5

Maximum and Minimum Properties of the Temperature Within the Dynamic Theory

§5.1. Maximum and Minimum Properties With Prescribed Heat Fluxes

The considerations of Chapter 2 tell us that, within the quasi-static theory of thermoelasticity which is obtained by omitting the inertial term $b\,\partial^2 u/\partial t^2$ from the thermoelastic equation (1.1.2), the temperature does not satisfy the classical Maximum Principle. The same is true, with even greater force, of the dynamic theory, that is the theory which results when the inertial term is retained and the full thermoelastic equations (1.1.1) and (1.1.2) are used to determine the temperature.

This fact notwithstanding, it is the case that the temperature exhibits maximum and minimum properties provided that the temperatures or the heat fluxes at the boundaries are suitably restricted, and provided we consider the behaviour of the temperature at values of t which are sufficiently large for the transient disturbance generated by the initial data on $t = 0$ to have died down.

In this section we prove a theorem† which confirms the correctness of our assertion when the thermal boundary conditions are

$$\left.\frac{\partial \theta}{\partial x}\right|_{x=0} = f, \qquad \left.\frac{\partial \theta}{\partial x}\right|_{x=1} = g, \tag{5.1.1}$$

† See [20].

and, as ever, the faces are held fixed, that is to say the condition (1.1.3) holds.

The *net heat flux* is the difference

$$h = g - f.$$

Thus, there is a net flux of heat into the body at any instant t at which $h(t) > 0$, and there is a net flux of heat out of the body at any instant at which $h(t) < 0$.

If I is any compact rectangle of the form $[0, 1] \times [t_1, t_2]$ we shall use the notation dI for the *parabolic boundary* of I, that is for the union of the line segments

$$[0, 1] \times \{t_1\}, \qquad \{0\} \times [t_1, t_2], \qquad \{1\} \times [t_1, t_2],$$

which comprise three sides of the rectangle (the uppermost part of the boundary being omitted).

We shall say that θ *has the maximum property* in a half-strip $[0, 1] \times [t_0, \infty)$ if

$$\max_I \theta = \max_{dI} \theta,$$

whenever I is a compact rectangle of the form $[0, 1] \times [t_1, t_2]$, with $t_1 \geqq t_0$.

In the same way, we shall say that θ *has the minimum property* in a half-strip $[0, 1] \times [t_0, \infty)$ if

$$\min_I \theta = \min_{dI} \theta,$$

whenever I is a compact rectangle of the form $[0, 1] \times [t_1, t_2]$, with $t_1 \geqq t_0$.

Theorem 19. *Suppose that θ and u are continuous in $[0, 1] \times [0, \infty)$ and C^4 in $[0, 1] \times (0, \infty)$ and that they satisfy the thermoelastic equations (1.1.1) and (1.1.2) and the boundary conditions (5.1.1) and (1.1.3). Suppose, in addition, that as $t \to \infty$ and $T \to \infty$,*

(i) $h(t) \to \infty$ *or* $h(t) \to -\infty$,

(ii) $h'(t) = o(|h(t)|)$,

(iii) $\displaystyle\int_0^T ((f''(t))^2 + (g''(t))^2)\, dt = o((h(T))^2).$

Then there is a finite $t_0 \geqq 0$ such that θ has the $\begin{Bmatrix} \text{maximum} \\ \text{minimum} \end{Bmatrix}$ property in $[0, 1] \times [t_0, \infty)$ according as $h(t) \to \begin{Bmatrix} -\infty \\ \infty \end{Bmatrix}$.

The hypotheses (i), (ii), (iii) are satisfied if, for example, f, g, and h $(=g-f)$, are all polynomials in t of the same degree $\geqq 1$.

We might regard (iii) as saying that when we calculate the net heat flux h, by subtracting f from g, the amount of cancellation is not excessive.

If one of the faces is *thermally insulated*, that is if $f = 0$ or $g = 0$, (iii) reduces to the condition

(iv) $\displaystyle\int_0^T (h''(t))^2\,dt = o((h(T))^2)$

on the heat flux. In that case, (i), (ii), and (iv) are all satisfied if h is any nonconstant polynomial in t.

In order to prove Theorem 19 we begin by observing that if we integrate both sides of the thermoelastic equation (1.1.1) with respect to x over the unit interval $[0, 1]$ and appeal to the boundary conditions we deduce the formula

$$h = \frac{d}{dt}\int_0^1 \theta\,dx \tag{5.1.2}$$

for the net heat flux.

Next, we integrate the second thermoelastic equation (1.1.2) with respect to x, and use the boundary condition (1.1.3), to obtain the equation

$$\begin{aligned} u(x, t) = &-(1-x)\int_0^x y\left(\sqrt{a}\cdot\frac{\partial\theta}{\partial y}(y, t) + b\,\frac{\partial^2 u}{\partial t^2}(y, t)\right)dy \\ &- x\int_x^1 (1-y)\left(\sqrt{a}\cdot\frac{\partial\theta}{\partial y}(y, t) + b\,\frac{\partial^2 u}{\partial t^2}(y, t)\right)dy. \end{aligned}$$

On differentiating with respect to x and applying partial integration to the integrals which involve θ we find the strain to be

$$\begin{aligned} \frac{\partial u}{\partial x}(x, t) = &\sqrt{a}\cdot\theta(x, t) - \sqrt{a}\int_0^1 \theta(y, t)\,dy \\ &+ b\int_0^x y\,\frac{\partial^2 u}{\partial t^2}(y, t)\,dy - b\int_x^1 (1-y)\,\frac{\partial^2 u}{\partial t^2}(y, t)\,dy \end{aligned}$$

and hence, we find the strain rate to be

$$\begin{aligned} \frac{\partial^2 u}{\partial x\,\partial t}(x, t) = &\sqrt{a}\cdot\frac{\partial\theta}{\partial t}(x, t) - \sqrt{a}\cdot h(t) \\ &+ b\int_0^x y\,\frac{\partial^3 u}{\partial t^3}(y, t)\,dy - b\int_x^1 (1-y)\,\frac{\partial^3 u}{\partial t^3}(y, t)\,dy. \end{aligned}$$

When we substitute this last expression back into the right-hand side of (1.1.1) we conclude that

$$\frac{\partial^2\theta}{\partial x^2}(x, t) = (1 + a)\frac{\partial\theta}{\partial t}(x, t) - ah(t)$$
$$+ b\sqrt{a}\int_0^x y\frac{\partial^3 u}{\partial t^3}(y, t)\,dy - b\sqrt{a}\int_x^1 (1 - y)\frac{\partial^3 u}{\partial t^3}(y, t)\,dy.$$

Thus

$$\left|\frac{\partial^2\theta}{\partial x^2} - (1 + a)\frac{\partial\theta}{\partial t} + ah\right| \leqq b\sqrt{a}\int_0^1 \left|\frac{\partial^3 u}{\partial t^3}\right| dx$$
$$\leqq b\sqrt{a}\sqrt{\int_0^1 \left(\frac{\partial^3 u}{\partial t^3}\right)^2 dx} \leqq \sqrt{2abE_2}, \qquad (5.1.3)$$

where

$$E_2 = \frac{1}{2}\int_0^1 \left(\left(\frac{\partial^2\theta}{\partial t^2}\right)^2 + \left(\frac{\partial^3 u}{\partial x\,\partial t^2}\right)^2 + b\left(\frac{\partial^3 u}{\partial t^3}\right)^2\right) dx$$

is the higher-order total energy which results when we replace θ and u by their second derivatives with respect to t in the definition

$$E = \frac{1}{2}\int_0^1 \left(\theta^2 + \left(\frac{\partial u}{\partial x}\right)^2 + b\left(\frac{\partial u}{\partial t}\right)^2\right) dx.$$

According to Theorem 1, the derivative of E is given by the formula

$$E' + \int_0^1 \left(\frac{\partial\theta}{\partial x}\right)^2 dx = g\theta|_{x=1} - f\theta|_{x=0}$$

and, likewise, the derivative of E_2 is given by the formula

$$E_2' + \int_0^1 \left(\frac{\partial^3\theta}{\partial x\,\partial t^2}\right)^2 dx = g''\left.\frac{\partial^2\theta}{\partial t^2}\right|_{x=1} - f''\left.\frac{\partial^2\theta}{\partial t^2}\right|_{x=0}.$$

If now we integrate the identities

$$\frac{\partial}{\partial x}\left((1 - x)\frac{\partial^2\theta}{\partial t^2}\right) = -\frac{\partial^2\theta}{\partial t^2} + (1 - x)\frac{\partial^3\theta}{\partial x\,\partial t^2},$$
$$\frac{\partial}{\partial x}\left(x\frac{\partial^2\theta}{\partial t^2}\right) = \frac{\partial^2\theta}{\partial t^2} + x\frac{\partial^3\theta}{\partial x\,\partial t^2},$$

with respect to x over the unit interval and use (5.1.2) we deduce that

$$-\left.\frac{\partial^2\theta}{\partial t^2}\right|_{x=0} = -h' + \int_0^1 (1-x)\frac{\partial^3\theta}{\partial x\,\partial t^2}\,dx,$$

$$\left.\frac{\partial^2\theta}{\partial t^2}\right|_{x=1} = h' + \int_0^1 x\frac{\partial^3\theta}{\partial x\,\partial t^2}\,dx,$$

and, hence, we arrive at the equation

$$E_2' + \int_0^1 \left(\frac{\partial^3\theta}{\partial x\,\partial t^2}\right)^2 dx = h'h'' + f''\int_0^1 (1-x)\frac{\partial^3\theta}{\partial x\,\partial t^2}\,dx$$
$$+ g''\int_0^1 x\frac{\partial^3\theta}{\partial x\,\partial t^2}\,dx.$$

The sum of the second and third terms on the right-hand side does not exceed

$$|f''|\int_0^1 \left|\frac{\partial^3\theta}{\partial x\,\partial t^2}\right| dx + |g''|\int_0^1 \left|\frac{\partial^3\theta}{\partial x\,\partial t^2}\right| dx$$
$$\leqq \tfrac{1}{2}(f'')^2 + \tfrac{1}{2}(g'')^2 + \int_0^1 \left(\frac{\partial^3\theta}{\partial x\,\partial t^2}\right)^2 dx$$

and, therefore, E_2' does not exceed

$$h'h'' + \tfrac{1}{2}(f'')^2 + \tfrac{1}{2}(g'')^2.$$

Thus

$$E_2(T) \leqq E_2(0) - \tfrac{1}{2}(h'(0))^2 + \tfrac{1}{2}(h'(T))^2$$
$$+ \tfrac{1}{2}\int_0^T ((f''(t))^2 + (g''(t))^2)\,dt$$

for every positive T, and the hypotheses on f, g, h ensure that

$$E_2(t) = o((h(t))^2) \quad \text{as } t \to \infty.$$

On considering the alternative $h(t) \to \infty$ we see that, in that case, we can choose a finite $t_0 \geqq 0$ in such a way that

$$h(t) > 0 \quad \text{and} \quad \sqrt{2abE_2(t)} < ah(t), \qquad t > t_0.$$

In the light of the inequality (5.1.3), this choice ensures that the temperature satisfies the inequality

$$\frac{\partial^2\theta}{\partial x^2} < (1+a)\frac{\partial\theta}{\partial t}$$

in $[0, 1] \times [t_0, \infty)$ and, as a standard argument tells us, that θ has the minimum property in $[0, 1] \times [t_0, \infty)$.

The conclusion that θ has the maximum property in $[0, 1] \times [t_0, \infty)$ if $h(t) \to -\infty$ follows on making the obvious alterations.

§5.2. Maximum and Minimum Properties With Prescribed Temperatures

In this section we suppose once again that θ and u satisfy the thermoelastic equations, and that u vanishes on the faces, but we prescribe the temperatures

$$\theta|_{x=0} = f, \qquad \theta|_{x=1} = g, \tag{5.2.1}$$

of the faces rather than the heat fluxes.

We shall write $f \vee g$ and $f \wedge g$ for the greater and the lesser of the boundary temperatures, that is

$$f \vee g = \tfrac{1}{2}(f + g) + \tfrac{1}{2}|f - g|,$$
$$f \wedge g = \tfrac{1}{2}(f + g) - \tfrac{1}{2}|f - g|.$$

If the temperature were a solution of the heat equation it would follow that if both faces are heated then, to within an error which tends to zero as the time tends to infinity, the maximum temperature of the body would coincide with the greater of the temperatures of the faces, while if both faces are cooled the minimum temperature would coincide, again to within an error which tends to zero, with the lesser of the temperatures of the faces. In exact terms†:

Theorem 20. *Suppose that θ is continuous in $[0, 1] \times [0, \infty)$ and C^2 in $[0, 1] \times (0, \infty)$ and satisfies the heat equation*

$$\frac{\partial^2 \theta}{\partial x^2} = \frac{\partial \theta}{\partial t}$$

and the boundary conditions (5.2.1), *where f and g are C^1 in $[0, \infty)$ and both increase or both decrease for all large t, then*

$$\max\{\theta(x, t) \colon 0 \leqq x \leqq 1\} = (f \vee g)(t) + o(1) \quad \text{as } t \to \infty,$$

or

$$\min\{\theta(x, t) \colon 0 \leqq x \leqq 1\} = (f \wedge g)(t) + o(1) \quad \text{as } t \to \infty,$$

according as f and g increase or decrease for all large t.

---

† This result is proved in [15].

The conclusion, which is most readily established by taking advantage of the Maximum Principle, is valid independently of the initial values taken when $t = 0$.

When we attempt to pass from the heat equation to the thermoelastic equations we are forced to argue in a way which does not require that we have a Maximum Principle to hand. Our line of attack will be to construct a function ϕ which approximates θ in the uniform $o(1)$ sense as $t \to \infty$, and then to argue from the detailed behaviour of ϕ.

It will be necessary to impose the restriction $a < 2$ upon the coupling constant but this is likely to be amply satisfied in practice.

Since we need to consider derivatives of f and g of large orders, we cease to denote them by primes and use superscripts instead; thus $f^{(n)} = d^n f/dt^n$ and $f^{(0)} = f$.

Theorem 21. *Let $a < 2$, and suppose that θ and u are continuous in $[0, 1] \times [0, \infty)$ and C^3 in $[0, 1] \times (0, \infty)$ and satisfy the thermoelastic equations* (1.1.1) *and* (1.1.2) *and the boundary conditions* (1.1.3) *and* (5.2.1).

Suppose, in addition, that, for some integer $N \geqq 2$, f and g each have continuous derivatives of order $N + 3$, that $f^{(1)}$ and $g^{(1)}$ are both positive for all large t, or both are negative for all large t, that

$$f^{(n)}(t) = o(|f^{(1)}(t)|)$$

and
$$g^{(n)}(t) = o(|g^{(1)}(t)|) \quad \text{as } t \to \infty \qquad (n = 2, \ldots, N),$$

and that the derivatives

$$f^{(n)}, \quad g^{(n)} \qquad (n = N + 1, N + 2, N + 3)$$

are integrable and square-integrable on $[0, \infty)$.

Then

$$\max\{\theta(x, t)\colon 0 \leqq x \leqq 1\} = (f \vee g)(t) + o(1) \quad \text{as } t \to \infty,$$

or

$$\min\{\theta(x, t)\colon 0 \leqq x \leqq 1\} = (f \wedge g)(t) + o(1) \quad \text{as } t \to \infty,$$

according as $f^{(1)}$ and $g^{(1)}$ are both positive or both negative for all large t.

Once again, the theorem holds whatever the initial values taken by the temperature, the displacement, or the velocity.

The hypotheses upon f and g are much more restrictive than those needed to establish the corresponding result (Theorem 20) for the heat equation. They are satisfied, though, if f and g are nonconstant polynomials in t and both tend to ∞, or both tend to $-\infty$, as $t \to \infty$. In this case, of course, all the derivatives of f and g of a sufficiently high order vanish identically.

In order to construct a ϕ which approximates the temperature θ and a v which approximates the displacement u, we introduce two sequences $\{p_n\}_{n \geqq 0}$, $\{\pi_n\}_{n \geqq 0}$ of polynomials by requiring that they satisfy the differential equations

$$\frac{d^2 p_0}{dx^2} = 0,$$

$$\frac{d^2 p_n}{dx^2} = p_{n-1} + \sqrt{a} \cdot \frac{dp_{n-1}}{dx} \qquad (n \geqq 1),$$

$$\frac{d^2 \pi_0}{dx^2} = \sqrt{a} \cdot \frac{dp_0}{dx},$$

$$\frac{d^2 \pi_1}{dx^2} = \sqrt{a} \cdot \frac{dp_1}{dx},$$

$$\frac{d^2 \pi_n}{dx^2} = \sqrt{a} \cdot \frac{dp_n}{dx} + b\pi_{n-2} \qquad (n \geqq 2),$$

and the boundary conditions

$$p_0(0) = 0, \qquad p_0(1) = 1$$
$$p_n(0) = p_n(1) = 0 \qquad (n \geqq 1),$$
$$\pi_n(0) = \pi_n(1) = 0 \qquad (n \geqq 0).$$

When $a = b = 0$ the polynomials π_n reduce to zero identically, while the polynomials p_n reduce to a known set, namely the LIDSTONE *polynomials*.†

The polynomials become progressively more complicated as n increases, but it will suffice to know the explicit forms of just two of them: they are

$$p_0(x) = x,$$
$$p_1(x) = -\tfrac{1}{12}(2 - a)x - \tfrac{1}{4}ax^2 + \tfrac{1}{6}(1 + a)x^3.$$

† See, for example, BOAS and BUCK [3].

Since each of the polynomials p_n, apart from p_0, vanishes at $x = 0$ and at $x = 1$ there are polynomials P_n such that

$$p_n(x) = -x(1 - x)P_n(x) \qquad (n \geqq 1).$$

We notice that

$$P_1(x) = \tfrac{1}{12}(2 - a + 2(1 + a)x)$$

and that, for each x in $[0, 1]$,

$$P_1(x) \geqq \tfrac{1}{12}(2 - a) > 0$$

and

$$p_1(x) \leqq -\tfrac{1}{12}(2 - a)x(1 - x)$$

provided that $a < 2$.

If we introduce the constants

$$c_n = \max\{|P_n(x)| : 0 \leqq x \leqq 1\}$$

we have the estimates

$$|p_n(x)| \leqq c_n x(1 - x) \qquad (n \geqq 1),$$

which hold throughout the unit interval.

Our next step is to form the approximating sums

$$\phi(x, t) = \sum_{n=0}^{N} (p_n(1 - x)f^{(n)}(t) + p_n(x)g^{(n)}(t)),$$

$$v(x, t) = \sum_{n=0}^{N} (-\pi_n(1 - x)f^{(n)}(t) + \pi_n(x)g^{(n)}(t)),$$

in which N is the integer that occurs in the hypotheses of the theorem, and to introduce the differences

$$\psi = \theta - \phi, \qquad w = u - v.$$

We shall prove presently that

$$\psi \to 0 \quad \text{as } t \to \infty, \tag{5.2.2}$$

the convergence being uniform with respect to x in $[0, 1]$. If, momentarily, we take this fact for granted we can complete the proof in the following way. We deal only with the case in which $f^{(1)}$ and $g^{(1)}$ are both positive for all large t; the alternative case can be handled by making the obvious amendments.

Let $\varepsilon > 0$ be arbitrarily small and choose $t_0 > 0$ so that $|\psi| < \varepsilon$ in $[0, 1] \times (t_0, \infty)$ and $f^{(1)} > 0$ and $g^{(1)} > 0$ in (t_0, ∞). Since $p_0(x) = x$ we can write ϕ as

$$\begin{aligned}
\phi(x, t) &= (1 - x)f(t) + xg(t) + \sum_{n=1}^{N} (p_n(1 - x)f^{(n)}(t) + p_n(x)g^{(n)}(t)) \\
&= (1 - x)f(t) + xg(t) \\
&\quad + f^{(1)}(t)\left(p_1(1 - x) + \sum_{n=2}^{N} p_n(1 - x) \frac{f^{(n)}(t)}{f^{(1)}(t)}\right) \\
&\quad + g^{(1)}(t)\left(p_1(x) + \sum_{n=2}^{N} p_n(x) \frac{g^{(n)}(t)}{g^{(1)}(t)}\right) \\
&\leqq (1 - x)f(t) + xg(t) \\
&\quad + x(1 - x)f^{(1)}(t)\left(-\tfrac{1}{12}(2 - a) + \sum_{n=2}^{N} c_n \frac{|f^{(n)}(t)|}{f^{(1)}(t)}\right) \\
&\quad + x(1 - x)g^{(1)}(t)\left(-\tfrac{1}{12}(2 - a) + \sum_{n=2}^{N} c_n \frac{|g^{(n)}(t)|}{g^{(1)}(t)}\right),
\end{aligned}$$

and, since $f^{(n)} = o(f^{(1)})$ and $g^{(n)} = o(g^{(1)})$ when $n = 2, \ldots, N$, we can choose $t_1 > t_0$ in such a way that each of the sums

$$\sum_{n=2}^{N} c_n \frac{|f^{(n)}(t)|}{f^{(1)}(t)}, \qquad \sum_{n=2}^{N} c_n \frac{|g^{(n)}(t)|}{g^{(1)}(t)}$$

is strictly less than $(2 - a)/12$ whenever $t > t_1$. Thus the approximating ϕ satisfies

$$\phi(x, t) \leqq (1 - x)f(t) + xg(t)$$

in $[0, 1] \times (t_1, \infty)$ and, therefore, the temperature satisfies

$$\begin{aligned}
\theta(x, t) &= \phi(x, t) + \psi(x, t) \\
&< (1 - x)f(t) + xg(t) + \varepsilon \\
&\leqq (f \vee g)(t) + \varepsilon
\end{aligned}$$

in $[0, 1] \times (t_1, \infty)$. Thus

$$(f \vee g)(t) \leqq \max\{\theta(x, t): 0 \leqq x \leqq 1\} < (f \vee g)(t) + \varepsilon$$

in (t_1, ∞), and we have arrived at the desired conclusion, namely

$$\max\{\theta(x, t): 0 \leqq x \leqq 1\} = (f \vee g)(t) + o(1).$$

It remains for us to verify (5.2.2), that is that ψ tends to zero uniformly. We start by asking how close the approximating functions ϕ and v come to satisfying the thermoelastic equations. If we appeal to the equations which define the sequences $\{p_n\}_{n\geqq 0}$, $\{\pi_n\}_{n\geqq 0}$ we find, after some calculation, that

$$\begin{aligned}\frac{\partial^2\phi}{\partial x^2}-\frac{\partial\phi}{\partial t}-\sqrt{a}\cdot\frac{\partial^2 v}{\partial x\,\partial t}&=-(p_N(1-x)+\sqrt{a}\cdot\pi_N'(1-x))f^{(N+1)}(t)\\&\quad-(p_N(x)+\sqrt{a}\cdot\pi_N'(x))g^{(N+1)}(t)\\&=-\zeta(x,t),\quad\text{say,}\end{aligned}$$

and that

$$\begin{aligned}\frac{\partial^2 v}{\partial x^2}-\sqrt{a}\cdot\frac{\partial\phi}{\partial x}-b\frac{\partial^2 v}{\partial t^2}&=b\pi_{N-1}(1-x)f^{(N+1)}(t)-b\pi_{N-1}(x)g^{(N+1)}(t)\\&\quad+b\pi_N(1-x)f^{(N+2)}(t)-b\pi_N(x)g^{(N+2)}(t)\\&=-\xi(x,t),\quad\text{say.}\end{aligned}$$

Moreover, ϕ and v satisfy the same boundary conditions as do θ and u, namely

$$\phi|_{x=0}=f,\qquad \phi|_{x=1}=g,$$
$$v|_{x=0}=v|_{x=1}=0,$$

and, therefore, the differences $\psi=\theta-\phi$ and $w=u-v$ satisfy the inhomogeneous thermoelastic equations

$$\frac{\partial^2\psi}{\partial x^2}=\frac{\partial\psi}{\partial t}+\sqrt{a}\cdot\frac{\partial^2 w}{\partial x\,\partial t}+\zeta,$$

$$\frac{\partial^2 w}{\partial x^2}=\sqrt{a}\cdot\frac{\partial\psi}{\partial x}+b\frac{\partial^2 w}{\partial t^2}+\xi,$$

and the homogeneous boundary conditions

$$\psi|_{x=0}=\psi|_{x=1}=w|_{x=0}=w|_{x=1}=0.$$

The hypotheses of integrability and square-integrability imposed upon the derivatives $f^{(n)}$ and $g^{(n)}$, for $n=N+1$, $N+2$, $N+3$, ensure the convergence of the integrals

$$\int_0^\infty\int_0^1\zeta^2\,dx\,dt,\qquad \int_0^\infty\left(\sqrt{\int_0^1\xi^2\,dx}\right)dt,$$

$$\int_0^\infty\int_0^1\left(\frac{\partial\zeta}{\partial t}\right)^2dx\,dt,\qquad \int_0^\infty\left(\sqrt{\int_0^1\left(\frac{\partial\xi}{\partial t}\right)^2dx}\right)dt,$$

and we shall use this fact, together with energy integral arguments, to deduce the uniform convergence of ψ to zero.

It will be enough to establish the convergence of the integrals

$$\int_0^\infty \int_0^1 \left(\frac{\partial\psi}{\partial x}\right)^2 dx\, dt, \qquad \int_0^\infty \int_0^1 \left(\frac{\partial^2\psi}{\partial x\, \partial t}\right)^2 dx\, dt.$$

Let us put

$$E = \tfrac{1}{2}\int_0^1 \left(\psi^2 + \left(\frac{\partial w}{\partial x}\right)^2 + b\left(\frac{\partial w}{\partial t}\right)^2\right) dx,$$

$$F = \int_0^1 \left(\frac{\partial\psi}{\partial x}\right)^2 dx,$$

$$G = \int_0^1 \zeta^2\, dx,$$

$$H = \sqrt{\frac{8}{b}\int_0^1 \xi^2\, dx},$$

so that E is the total energy associated with the temperature difference ψ and the displacement difference w, and G and H are known to be integrable on $[0, \infty)$. We seek to prove that F too is integrable.

A slight variant of the argument used to prove Theorem 1 leads us to conclude that the derivative of E is given by the formula

$$E' + F = -\int_0^1 \left(\zeta\psi + \xi\frac{\partial w}{\partial t}\right) dx.$$

Since

$$\begin{aligned} -\int_0^1 \zeta\psi\, dx &\leqq \tfrac{1}{2}\int_0^1 (\psi^2 + \zeta^2)\, dx \\ &\leqq \tfrac{1}{2}\int_0^1 \left(\frac{\partial\psi}{\partial x}\right)^2 dx + \tfrac{1}{2}\int_0^1 \zeta^2\, dx \\ &= \tfrac{1}{2}(F + G) \end{aligned}$$

and

$$-\int_0^1 \xi\frac{\partial w}{\partial t}\, dx \leqq \sqrt{\frac{2}{b}\int_0^1 \xi^2\, dx}\,\sqrt{\frac{b}{2}\int_0^1 \left(\frac{\partial w}{\partial t}\right)^2 dx} \leqq \tfrac{1}{2}H\sqrt{E},$$

we have deduced the differential inequality

$$2E' + F \leqq G + H\sqrt{E}. \tag{5.2.3}$$

Because F is nonnegative the latter inequality implies the weaker inequality

$$2E' \leqq G + H\sqrt{E}$$

from which we can deduce that E is bounded on $[0, \infty)$. For, let T be any positive number, and let $m(T)$ be the maximum value attained by E on the interval $[0, T]$. Integration of the weaker inequality leads us to conclude, as G and H are integrable, that on $[0, T]$

$$\begin{aligned} 2E(t) &\leqq 2E(0) + \int_0^t G(s)\,ds + \int_0^t H(s)\sqrt{E(s)}\,ds \\ &\leqq 2E(0) + \int_0^\infty G(s)\,ds + \sqrt{m(T)}\int_0^\infty H(s)\,ds \end{aligned}$$

and, therefore, that

$$\begin{aligned} 2m(T) &\leqq 2E(0) + \int_0^\infty G(s)\,ds + \sqrt{m(T)}\int_0^\infty H(s)\,ds \\ &\leqq 2E(0) + \int_0^\infty G(s)\,ds + m(T) + \tfrac{1}{4}\left(\int_0^\infty H(s)\,ds\right)^2 \end{aligned}$$

or, in other words,

$$m(T) \leqq 2E(0) + \int_0^\infty G(s)\,ds + \tfrac{1}{4}\left(\int_0^\infty H(s)\,ds\right)^2.$$

Since T is arbitrary it follows that E is bounded on the entire interval $[0, \infty)$.

On returning to the stronger differential inequality (5.2.3) we see that

$$2E' + F \leqq G + \sqrt{\sup E} \cdot H,$$

and, when we integrate with respect to t over $[0, T]$ and omit a term $2E(T)$ from the left-hand side, we deduce that

$$\begin{aligned} \int_0^T F(t)\,dt &\leqq 2E(0) + \int_0^T G(t)\,dt + \sqrt{\sup E}\int_0^T H(t)\,dt \\ &\leqq 2E(0) + \int_0^\infty G(t)\,dt + \sqrt{\sup E}\int_0^\infty H(t)\,dt \end{aligned}$$

for every positive T. Thus F is integrable and, therefore, the integral

$$\int_0^\infty \int_0^1 \left(\frac{\partial \psi}{\partial x}\right)^2 dx\, dt$$

converges.

Finally, we differentiate, with respect to t, the inhomogeneous thermoelastic equations and the homogeneous boundary conditions that ψ and w satisfy, and we conclude in the same way that the integral

$$\int_0^\infty \int_0^1 \left(\frac{\partial^2 \psi}{\partial x\, \partial t}\right)^2 dx\, dt$$

converges; this completes the proof.

References

[1] BELLMAN, R. A property of summation kernels. *Duke Math. J.* **15**, 1013–1019 (1948).

[2] BESICOVITCH, A. S. *Almost Periodic Functions.* Cambridge, Cambridge University Press, 1932.

[3] BOAS, R. P. and R. C. BUCK. *Polynomial Expansions of Analytic Functions.* Berlin, Springer, 1964.

[4] BOLEY, B. A. and J. H. WEINER. *Theory of Thermal Stresses.* New York, Wiley, 1960.

[5] CANNON, J. R. The one-dimensional heat equation. In *Encyclopedia of Mathematics and its Applications*, Vol. 23, Reading, Mass., Addison-Wesley, 1984.

[6] CARLSON, D. E. Linear thermoelasticity. In *Handbuch der Physik.* Bd. **Vla/2**, edited by C. Truesdell. Berlin, Springer, 1972.

[7] CARSLAW, H. S. and J. C. JAEGER. *Conduction of Heat in Solids*, 2nd edition. Oxford, Clarendon Press, 1959.

[8] CHADWICK, P. Thermoelasticity. The dynamical theory. In *Progress in Solid Mechanics*, Vol. I. Amsterdam, North-Holland, 1960.

[9] COLEMAN, B. D., M. FABRIZIO, and D. R. OWEN. On the thermodynamics of second sound in dielectric crystals. *Arch. Rational Mech. Anal.* **80**, 135–158 (1982).

[10] DAY, W. A. Justification of the uncoupled and quasi-static approximations in a problem of dynamic thermoelasticity. *Arch. Rational Mech. Anal.* **77**, 387–396 (1981).

[11] DAY, W. A. Further justification of the uncoupled and quasi-static approximations in thermoelasticity. *Arch. Rational Mech. Anal.* **79**, 85–95 (1982).

[12] DAY, W. A. Extensions of a property of the heat equation to linear thermoelasticity and other theories. *Quart. Appl. Math.* **40**, 319–330 (1982).

[13] DAY, W. A. A decreasing property of solutions of parabolic equations with applications to thermoelasticity. *Quart. Appl. Math.* **41**, 468–475 (1983).

[14] DAY, W. A. Mean and recurrence properties of the temperature in dynamic thermoelasticity. *J. Elasticity* **13**, 225–230 (1983).

[15] DAY, W. A. A property of the heat equation which extends to the thermoelastic equations. *Arch. Rational Mech. Anal.* **83**, 99–113 (1983).

[16] DAY, W. A. Steady state oscillatory temperatures in coupled, quasi-static thermoelasticity. *Quart. J. Mech. Appl. Math.* **37**, 581–596 (1984).

[17] DAY, W. A. A comment on approximations to the temperature in dynamic linear thermoelasticity. *Arch. Rational Mech. Anal.* **85**, 237–250 (1984).

[18] DAY, W. A. On the failure of the Maximum Principle in coupled thermoelasticity. *Arch. Rational Mech. Anal.* **86**, 1–12 (1984).

[19] DAY, W. A. Initial sensitivity to the boundary in coupled thermoelasticity. *Arch. Rational Mech. Anal.* **87**, 253–266 (1985).

[20] DAY, W. A. Maximum and minimum properties of the temperature in linear thermoelasticity. *Quart. Appl. Math.* (to appear).

[21] FOURIER, J-B. J. *Théorie Analytique de la Chaleur.* Paris, Didot, 1822.

[22] FRIEDMAN, A. Monotonic decay of solutions of parabolic equations with nonlocal boundary conditions. *Quart. Appl. Math.* (to appear).

[23] GREEN, J. W. *Aspects of the Calculus of Variations.* Notes after lectures by Hans Lewy. Berkeley, University of California Press, 1939.

[24] HARDY, G. H., J. E. LITTLEWOOD, and G. PÓLYA. *Inequalities*, 2nd edition. Cambridge, Cambridge University Press, 1952.

[25] KAC, M. On some connections between probability theory and differential and integral equations. *Proceedings of the Second Berkeley Symposium on Mathematical Statistics and Probability.* Berkeley, University of California Press, 1951.

[26] KAC, M. Can one hear the shape of a drum? *Amer. Math. Monthly* **73**, 1–23 (1966).

[27] LEVI, E. E. Sull' equazione del calore. *Reale Accad. dei Lincei, Roma. Rendiconti* (5) **162**, 450–456 (1907).

[28] LEVI, E. E. Sull' equazione del calore. *Annali di Mat. Pura ed. Appl.* **14**, 187–264 (1908).

[29] NOWACKI, W. *Thermoelasticity.* Reading, Mass., Addison-Wesley, 1962.

[30] PÓLYA, G. and G. SZEGÖ. Sur quelques propriétés qualitatives de la propagation de la chaleur. *C.R. Acad. Sci. (Paris)* **192**, 1340–1342 (1931).

[31] PROTTER, M. H. and H. F. WEINBERGER. *Maximum Principles in Differential Equations.* Englewood Cliffs, N.J., Prentice-Hall, 1967.

[32] WIDDER, D. V. *The Heat Equation.* New York, Academic Press, 1975.

[33] WIENER, N. Generalized harmonic analysis. *Acta. Math.* **55**, 117–258 (1930).

[34] WIENER, N. *The Fourier Integral and Certain of Its Applications.* Cambridge, Cambridge University Press, 1933.

Index

Vol. 19 Knops, Payne: Uniqueness Theorems in Linear Elasticity
IX, 130 pages. 1971.

Vol. 20 Edelen, Wilson: Relativity and the Question of Discretization in Astronomy
With 34 figures. XII, 186 pages. 1970.

Vol. 21 McBride: Obtaining Generating Functions
XIII, 100 pages. 1971.

Vol. 22 Day: The Thermodynamics of Simple Materials with Fading Memory
With 8 figures. X, 134 pages. 1972.

Vol. 23 Stetter: Analysis of Discretization Methods for Ordinary Differential Equations
With 12 figures. XVI, 388 pages. 1973.

Vol. 24 Strieder/Aris: Variational Methods Applied to Problems of Diffusion and Reaction
With 12 figures. IX, 109 pages. 1973.

Vol. 25 Bohl: Momotonie: Lösbarkeit und Numerik bei Operatorgleichungen
Mit 9 Abbildungen. IX, 255 Seiten. 1974.

Vol. 26 Romanov: Integral Geometry and Inverse Problems for Hyperbolic Equations
With 21 figures, VI, 152 pages. 1974.

Vol. 27 Joseph: Stability of Fluid Motions I
With 57 figures. XIII, 282 pages. 1976.

Vol. 28 Joseph: Stability of Fluid Motions II
With 39 figures. XIV, 274 pages. 1976.

Vol. 29 Bressan: Relativistic Theories of Materials
XIV, 290 pages. 1978.

Vol. 30 Day: Heat Conduction within Linear Thermoelasticity
VII, 82 pages. 1985.